Basic Calculations for Chemical and Biological Analyses

Bassey J. S. Efiok, Ph.D.
National Heart Lung Blood Institute
National Institutes of Health
Bethesda, MD, USA

By AOAC INTERNATIONAL
481 N. Frederick Avenue, Suite 500
Gaithersburg, MD 20877-2504 USA

Printed in the United States of America
Second printing 1994
Third printing 1995

ISBN 0-935584-51-X

Dedication

Dedicated to the memory of my grandfather, Stephen Udo Akpan, whose exemplary wisdom, courage, and leadership still inspire me to strive for higher goals.

Table of Contents

Quick Reference to Abbreviations and Symbols*

A	absorbance
at. wt	atomic weight
a	ionic activity
bp	base pairs
c	concentration
Ci	curie
concn, C	concentration
D	Dalton
dpm	disintegrations per minute
E	extinction coefficient
e	base of natural logarithm
equiv	equivalent
equiv wt	equivalent weight
fw	formula weight
g	gram
HOAc	acetic acid
I	Ionic strength
IU	international unit
k	rate constant
L	liter
l	length of light path
log	logarithm to the base 10
M	molarity
mol	mole
mol wt	molecular weight
N	Normal (concentration)
N	Avogadro's number
n	number of equivalents per mole
OAc^-	acetate ion
P	product (of enzyme reaction)
S	substrate (for enzyme)
sp act.	specific activity
sp gr	specific gravity
t	time
V, vol	volume
wt	weight
w/w	weight-to-weight ratio
w/v	weight-to-volume ratio
Z	number of charge on ion
γ	activity coefficient
ρ	density
ϵ	molar extinction coefficient
μ	micro
[ ]	molar concentration of substance in bracket

Quick Reference to Unit Conversion*

Gram (g)

Unit	Symbol	Equivalent (g)
picogram	pg	1 pg = 10^{-12}
nanogram	ng	1 ng = 10^{-9}
microgram	µg	1 µg = 10^{-6}
milligram	mg	1 mg = 10^{-3}
kilogram	kg	1 kg = 10^{3}

Mole (mol)

Unit	Symbol	Equivalent (mol)
picomole	pmol	1 pmol = 10^{-12}
nanomole	nmol	1 nmol = 10^{-9}
micromole	µmol	1 µmol = 10^{-6}
millimole	mmol	1 mmol = 10^{-3}

Liter (l or L)

Unit	Symbol	Equivalent (L)
nanoliter	nL	1 nL = 10^{-9}
microliter	µL	1 µL = 10^{-6}
milliliter	mL	1 mL = 10^{-3}

Curie (Ci)

Unit	Symbol	Equivalent
curie	Ci	1 Ci = 2.22×10^{-12} dpm
microcurie	µCi	1 µCi = 10^{-6} Ci
millicurie	mCi	1 mCi = 10^{-3} Ci

Note: cpm = dpm × % efficiency

*See Appendix B and Table 5.1 for detailed lists.

Quick Reference to Selected Equations

Equations for Calculating Amounts and Concentrations of Reagents (See guide to abbreviations in page v)

$$\text{mol} = \frac{\text{wt (g)}}{\text{mol wt}}$$

$$\text{equiv} = \frac{\text{wt (g)}}{\text{equiv wt}} \quad \text{or} \quad \text{equiv} = \text{mol} \times n$$

$$\text{M} = \frac{\text{mol}}{\text{L}} \quad \text{or} \quad \text{M} = \frac{\text{wt}}{\text{mol wt} \times \text{L}}$$

$$\text{N} = \frac{\text{equiv}}{\text{L}} \quad \text{or} \quad \text{N} = \frac{\text{wt}}{\text{equiv wt} \times \text{L}}$$

Also,

$$\text{N} = n\text{M}$$

$$\rho = \frac{\text{wt}}{\text{vol}}$$

$$\text{sp gr} = \frac{\rho_{\text{sample}}}{\rho_{\text{H}_2\text{O}}} = \rho_{\text{sample}} \text{ at } 4°\text{C}$$

If a volume (V_1) of a reagent (concn C_1) is added to a diluent to yield a final volume (V_2 and a final concentration (C_2), then,

$$\%\ (\text{w/w}) = \frac{\text{g of analyte in sample}}{\text{g of sample}} \times 100\%$$

$$\%\ (\text{w/v}) = \frac{\text{g of analyte in sample}}{\text{mL of sample}} \times 100\%$$

$$\text{ppm (w/w)} = \frac{\text{g of analyte in sample}}{\text{g of sample}} \times 10^6 \text{ ppm}$$

$$\text{ppm (w/v)} = \frac{\text{g of analyte in sample}}{\text{mL of sample}} \times 10^6 \text{ ppm}$$

$$\text{Dilution factor} = \frac{C_1}{V_2} = \frac{V_2}{C_1}$$

$$C_1V_1 = C_2V_2$$

$$V_1 = \frac{C_2V_2}{C_1}$$

Equations for Calculating the Weight or Volume of a Reagent Needed to Prepare a Given Solution

For molar concentration,

$$\text{wt(g)} = \text{mol wt (g)} \times \text{M} \times \text{L}$$

When correcting for % purity,

$$\text{wt(g)} = \text{mol wt (g)} \times \text{M} \times \text{L} \times \frac{100}{\text{purity of reagent (\%)}}$$

For % and ppm concentration,

$$\text{wt(g)} = \frac{\text{\% (w/v)} \times \text{mL of solution}}{100\%}$$

$$\text{wt(g)} = \frac{\text{ppm (w/v)} \times \text{mL of solution}}{10^{-6}\ \text{ppm}}$$

For liquid reagents,

$$V_1 = \frac{C_2V_2}{C_1}$$

where V_1 is the volume of the stock reagent which, when diluted to a final volume V_2, yields a desired concentration C_2; C_1 is the concentration of the stock reagent. The units of C_1 and C_2 must be identical, and those of V_1 and V_2 must also be identical.

Equations for Enzyme Activity, Units, and Kinetic Constants

$$\text{activity} = \frac{\text{amount of substrate converted}}{t}$$

or

$$\text{activity} = \frac{\text{amount of product liberated}}{t}$$

$$\text{sp act.} = \frac{\text{amount of substrate converted}}{t \times \text{mg of protein in assay}}$$

or

$$\text{sp act.} = \frac{\text{amount of product liberated}}{t \times \text{mg of protein in assay}}$$

$$1\ \text{IU} = 1\ \mu\text{mol/min}$$

$$\text{Turnover number} = \frac{\text{mol of substrate catalyzed}}{\text{mol of enzyme} \times \text{s}}$$

In a Lineweaver-Burke plot,

$$V_{max} = \frac{1}{\text{y intercept}}$$

$$K_m = -\frac{1}{\text{x intercept}}$$

$$\frac{K_m}{V_{max}} = \text{slope}$$

OTHER COMMONLY USED EQUATIONS

Beer-Lambert Equation

$$A = E\ell C$$

Radioactivity remaining [N(*t*)] after time (*t*) has elapsed

$$N_t = N_0 e^{-(0.693t)/t_{1/2}}$$

Ionic strength

$$I = \tfrac{1}{2} \Sigma M_i Z_i^2$$

Preface

Basic Calculations for Chemical and Biological Analyses

Biochemical calculations are usually needed at some stage of a laboratory experiment. The types of calculations may vary but often include basic tasks such as: calculating the amount of a reagent needed to prepare a given solution or yield a given concentration in an assay medium; converting raw experimental data into concentrations, amounts, and activities; interconverting units of measure; and calculating variables or constants for biochemical reactions.

Students and technicians typically learn the skills to perform these calculations from their general or analytical chemistry courses. However, many soon forget how to apply the principles taught in these courses, or develop complex techniques that do not always yield correct results. I have helped many students solve different quantitative problems during my graduate and postdoctoral years at The Illinois Institute of Technology. This experience and personal communication with other students and technicians indicate the need for a book to closely guide students and technicians everywhere in understanding and applying basic biochemical calculations using simple, consistent techniques. This book is written to fill this need.

The topics covered in the five main chapters have wide applications in life science and chemical laboratories. They include: reagent concentrations, amounts, and preparation; acid-base, buffer, and pH calculations, including techniques for buffer preparation; spectrophotometry and its quantitative applications; and the quantitative aspects of radioactivity. Each chapter includes: (1) description and definition of basic principles fundamental to understanding how subsequent equations are derived and applied, (2) units of measure and their interconversion; (3) related chemical and/or biochemical techniques and their applications; and (4) several practical examples that illustrate how the basic equations are used to solve a wide range of common quantitative problems in analytical and research laboratories. A distinguishing feature of this book is that most of the examples are real problems. Even the few devised examples have been carefully constructed to closely approximate real problems that the reader will easily recognize and apply.

To assure that the main text is easy to read and understand, the derivation of some fundamental equations were placed in an appendix. Relevant materials, such as instrumentation, principles of measurement, physical/chemical constants, SI units tables, and recipes for preparing common laboratory buffers, are also included in the appendices.

Finally, quick reference guides to abbreviations, SI units, and equations, and an index to the practical examples are provided.

I would like to express my sincere gratitude to Professor Dale A. Webster of the Department of Biology, Illinois Institute of Technology, Chicago; Dr. Patricia A. Cunniff, Professor of Chemistry, Prince George's Community College; and Dr. Peter Spare of the McKellar General Hospital, Department of Chemistry Laboratories, Ontario, Canada. Their efforts helped in making this book simple to read and understand. May I also thank Krystyna A. McIver, Director of Publications, and other staff of AOAC INTERNATIONAL for their assistance and patience.

To all users of this book, I will sincerely appreciate your comments and suggestions.

Bassey J.S. Efiok, Ph.D
Molecular Hematology Branch
NHLBI, National Institutes of Health
Building 10, Room 7D-18
Bethesda, MD 20892

Chapter 1

REAGENT QUANTITATION: CALCULATING AMOUNTS AND CONCENTRATIONS; PREPARATION GUIDELINES

This chapter reviews the basic chemical concepts and units used to quantitate reagents (*1*, *2*). Equations for calculating the amounts and concentrations of reagents are derived immediately after the relevant chemical concept has been explained. A guideline on reagent preparation is then given, and, finally, practical examples addressing various aspects of calculating for reagent preparation, concentrations, and amounts are presented.

1.1. Review of Basic Atomic and Molecular Concepts

A. ATOMIC MASS UNIT (amu)

By international agreement, 1 amu is defined as $^1/_{12}$ the mass of an atom of carbon-12 (^{12}C). When converted to grams, 1 amu = 1.66×10^{-24} g.

B. MOLE (mol)

Weighing out or counting individual atoms or molecules is impossible because they are extremely small. The unit of measure, the mole, was introduced to enable scientists to weigh out the amount of any chemical substance that contains a given number of atoms, molecules, or other chemical particles. One mole of atoms is the number of atoms in exactly 12 g of ^{12}C, which is 6.02×10^{23}. In a broader sense, 1 mole of any substance contains 6.02×10^{23} units. One mole of a chemical element or compound contains 6.02×10^{23} atoms or molecules, respectively, and its weight in grams is equal to the numerical value of the **atomic weight (at. wt)** or **molecular weight (mol wt)** (see sections D and F). For example, the atomic weight of calcium is 40.08 amu. One mole of calcium weighs 40.08 g and contains 6.02×10^{23} atoms. It follows that **the number of moles in a given weight of a substance can be calculated using equation 1, and the weight of a given amount of moles can be calculated using equation 2, below:**

By definition,

$$\text{mol} = \frac{\text{wt(g)}}{\text{mol wt (g/mol)}} \tag{1}$$

$$\therefore \text{ wt (g)} = \text{mol} \times \text{mol wt (g/mol)} \tag{2}$$

If the substance is an element, molecular weight is substituted with atomic weight.

Examples: See problems 1, 2a, 5, and 10b in section 1.6.

C. AVOGADRO'S NUMBER (N)

The constant, 6.02×10^{23} atoms/mole or 6.02×10^{23} molecules/mole (see above), is defined as Avogadro's number (N). Therefore, for an element,

$$N = 6.02 \times 10^{23} \text{ atoms} = 1 \text{ mol} = 1 \text{ g-at. wt}$$

For a compound,

$$N = 6.02 \times 10^{23} \text{ molecules} = 1 \text{ mol} = 1 \text{ g-mol wt}$$

Examples: See problems 3 and 4 in section 1.6.

D. ATOMIC WEIGHT (at. wt) AND GRAM-ATOMIC WEIGHT (g-at. wt)

Atomic weight is the weight (in amu) of one atom of an element. It comprises the weight of the protons and neutrons in the atomic nucleus. The gram-atomic weight of any element is the weight (in grams) of 6.02×10^{23} atoms or 1 mole of that element. The gram-atomic weight is numerically equal to the atomic weight. For example, the atomic weight of sulfur (S) is 32.06 amu; the g-at. wt is 32.06 g. The 32.06 g is calculated as follows: There are 6.022×10^{23} atoms/mole S. Therefore, the mass (amu) of 1 mole of S is

$$1 \text{ mol S} \times \frac{6.022 \times 10^{23} \text{ atoms S}}{1 \text{ mol S}} \times \frac{32.06 \text{ amu}}{1 \text{ atom S}}$$

$$= 1.9307 \times 10^{25} \text{ amu}$$

This is then converted from amu to grams:

$$1.9307 \times 10^{25} \text{ amu} \times \frac{1.6606 \times 10^{-24} \text{ g}}{1 \text{ amu}} = 32.06 \text{ g}$$

E. MOLECULAR FORMULA

The atoms that make up one molecule of a compound constitute the chemical or molecular formula of that compound. For example, the formula of a molecule of glucose, which consists of 6 atoms of carbon, 12 atoms of hydrogen, and 6 atoms of oxygen, is $C_6H_{12}O_6$. The atoms in this formula are covalently linked and do not dissociate in solution. Therefore, $C_6H_{12}O_6$ exists as a discrete molecule. In contrast, discrete molecules of some compounds exist only in concept because the molecules are formed by associated ions. For example, the formula of magnesium sulfate is $MgSO_4$. In solution, it dissociates into one magnesium ion (Mg^{2+}) and one sulfate ion (SO_4^{2-}). A discrete molecule of magnesium sulfate or any other dissociable compound, therefore, exists only in concept.

F. MOLECULAR WEIGHT (mol wt) AND GRAM-MOLECULAR WEIGHT (g-mol wt)

The molecular weight of a compound is the sum of the atomic weight (in amu) of all the atoms that make up one molecule of the compound. The molecular weight (in g) of any compound contains 6.02×10^{23} molecules or 1 mole of the compound. For this reason, the molecular weight, expressed in grams, is termed gram-molecular weight. The gram-molecular weight of a compound is numerically equal to its molecular weight, and this can be calculated similar to the example given in section D, above, for the g-at. wt of sulfur.

G. FORMULA WEIGHT (fw)

This unit is used in place of molecular weight to designate the weight of the formula of a compound that does not exist as a discrete molecule (section E). It is defined as the sum of the atomic weights of all the elements that comprise the chemical formula of the compound. The formula weight (g) of any compound contains 6.02×10^{23} molecules or 1 mole of the compound.

H. DALTON (D)

This unit is used to report the molecular or atomic mass of substances; 1 Dalton is equal to 1 amu, or $^{1}/_{12}$ the mass of one atom of ^{12}C (see section 1.2A). Although the number of Daltons for a molecule is equivalent to the molecular weight, the Dalton is generally reserved for reporting the masses of macromolecular substances such as proteins.

I. EQUIVALENT WEIGHT (equiv wt)

This is the weight of an acid or a base containing 1 mole of replaceable H^+ or OH^-, respectively, or the weight of a redox compound that contains 1 mole of exchangeable electrons, or the weight of an ionic substance carrying 1 mole of ions. Gram-equivalent weight is used when equivalent weight is expressed in grams. **The equivalent weight of a substance is calculated using equation 3, below:**

$$\text{equiv wt} = \frac{\text{mol wt}}{n} \qquad (3)$$

where n is the number of **equivalents** per mole (or the number of replaceable H^+ or OH^-, exchangeable electrons, or charge, per molecule or ion of the substance). When $n = 1$, then **equiv wt = mol wt**.

Examples: See problems 18 and 19 in section 1.6.

J. EQUIVALENT (equiv)

This is an operational term that defines the gram-equivalent weight of an acid, a base, an electron transferring substance, or an ion. One equivalent of a compound contains 1 g-equiv wt of the compound. **The number of equivalents in a given weight of a substance is calculated using equations 4 and 6, and the weight containing a given number of equivalents is calculated using equation 5, below.** By definition,

$$\text{equiv} = \frac{\text{wt (g)}}{\text{equiv wt (g)}} \tag{4}$$

$$\therefore \ \text{wt (g)} = \text{equiv} \times \text{equiv wt} \tag{5}$$

Another useful equation can be derived by combining equations 3 and 5:

$$\text{mol wt} = \frac{\text{wt}}{\text{mol}} = \text{equiv wt} \times n$$

or

$$\frac{\text{wt}}{\text{equiv wt}} = n \times \text{mol}$$

Since, by equation 4,

$$\text{equiv} = \frac{\text{wt}}{\text{equiv wt}}$$

it follows that:

$$\text{equiv} = n \times \text{mol} \tag{6}$$

Again, n is the number of equiv/mol.

Example: See problem 19 in section 1.6.

1.2. Calculating Concentrations Based on Mole and Equivalent

A. MOLARITY (M)

This concentration unit is defined as the number of moles of a substance per liter (**L**) of solution. For example, a 2 M solution of NaCl contains 2 moles of NaCl/L. **The molarity of a substance in solution is calculated using equations 7 and 8, and the weight of a substance needed to prepare a solution of a given volume (in liters) and molarity is calculated using equation 9, below.** By definition,

$$M = \frac{\text{mol}}{\text{L}} \tag{7}$$

By substituting equation 1 into equation 7,

$$M = \frac{\text{wt (g)}}{\text{mol wt} \times \text{L}} \tag{8}$$

$$\therefore \text{wt (g)} = M \times \text{mol wt} \times \text{L} \tag{9}$$

Examples: See problems 2b, 6, 7, 8, 10c, 11, 12, 13c, and 14 in section 1.6.

When the purity of a reagent is less than 100%, the calculated weight is corrected by multiplying with the factor:

$$\frac{100}{\text{purity of reagent (\%)}}$$

Equation 10 then becomes:

$$\text{wt (g)} = (\text{mol wt} \times M \times \text{L}) \times \frac{100}{\text{purity of reagent (\%)}} \tag{10}$$

Examples: See problems 12 and 14 in section 1.6.

B. MOLALITY

This is a concentration unit defined as the number of moles of a solute per kilogram of solvent; i.e., a 1 molal solution of glucose is prepared by adding 180.16 g (1 mol) of glucose to 1 kg of H_2O. **The molality of a substance in solution is calculated using equations 11 and 12, and the weight of a substance needed to prepare a solution with a given molality and solvent weight (in kg) is calculated using equation 13, below.** By definition,

$$\text{molality} = \frac{\text{mol}}{\text{kg of solvent}} \tag{11}$$

By substituting equation 1 into equation 11,

$$\text{molality} = \frac{\text{wt (g)}}{\text{mol wt} \times \text{kg of solvent}} \tag{12}$$

$$\therefore \text{ wt (g)} = \text{molality} \times \text{mol wt} \times \text{kg of solvent} \tag{13}$$

Example: See problem 9 in section 1.6.

C. OSMOLARITY (osM)

This concentration unit is defined as the moles of particles per liter of solution. For substances that do not dissociate in solution, the osmolar concentration is equal to the molar concentration. For ionizable substances, the osmolar concentration is equal to n times the molarity, where n is the number of ions produced per molecule upon ionization. Thus,

$$\text{osM} = n \times \text{M} \tag{14}$$

and

$$\text{milliosmolarity (mosM)} = n \times \text{mM} \tag{15}$$

where mM is 1000th of 1 M, and mosM is 1000th of 1 osM. Because osmolarity is too large a unit, milliosmolarity is more widely used.

D. NORMALITY (N)

This concentration unit is defined as the number of equivalents of a substance per liter of solution. For example, a 0.5 N solution of acetic acid contains 0.5 equiv/L. **Normality is calculated using equations 16 and 17, and the weight of a substance needed to prepare a solution of a given volume (in liters) and normality is calculated using equation 18, below:**

$$\text{N} = \frac{\text{equiv}}{\text{L}} \tag{16}$$

By substituting equation 4 into equation 16,

$$\text{N} = \frac{\text{wt (g)}}{\text{equiv wt (g)} \times \text{L}} \tag{17}$$

$$\text{wt (g)} = \text{N} \times \text{equiv wt (g)} \times \text{L} \tag{18}$$

Examples: See problems 20–22 in section 1.6.

E. MOLARITY RELATED TO NORMALITY

From equations 6 and 7,

$$\text{equiv} = n \times \text{mol}$$
$$\text{mol} = \text{M} \times \text{L}$$

where equiv and mol represent the number of equivalents and the number of moles, respectively.

Combining the two equations yields:

$$\text{equiv} = n \times \text{M} \times \text{L} \tag{19}$$

From equation 16,

$$\text{equiv} = \text{N} \times \text{L} \tag{20}$$

Combining equations 19 and 20 yields

$$\text{N} \times \text{L} = n \times \text{M} \times \text{L}$$
$$\therefore \text{N} = n\text{M} \tag{21}$$

where n is the number of equivalents per mole.

Examples: See problems 20 and 21 in section 1.6.

1.3. Calculating Concentrations Based on Weight

Weight-based concentration units in frequent use are percent (%), parts per million (ppm), parts per billion (ppb), density (ρ), specific gravity (sp gr), and specific volume (sp vol). The equations for calculating percent are derived from the following expression:

$$\%\ \text{analyte} = \times \frac{\text{wt of analyte in sample}}{\text{wt or vol of sample}} \times 100\% \tag{22}$$

where analyte refers to the substance being determined, and sample refers to the material that contains the analyte and is being analyzed. If the **amount of sample** (denominator) is in a weight unit, the calculated **percent analyte** is designated % (w/w); if it is in a volume unit, the percent analyte is designated % (w/v). To derive similar expressions for calculating parts per million and parts per billion, the 100% is replaced by 10^6 ppm or 10^9 ppb, respectively.

Specific equations for calculating each of the weight-based concentrations, and a corresponding weight of reagent needed to prepare a solution of a given concentration are derived below.[1]

A. PERCENT (WEIGHT/WEIGHT), or % (w/w)

This is the grams of pure analyte in 100 g of sample. For example, a 37% (w/w) solution of commercial concentrated HCl contains 37 g of pure HCl in 100 g of solution. **The % (w/w) of an analyte in a sample is calculated using equation 23 below:**

$$\% \text{ (w/w)} = \frac{\text{g of analyte in sample}}{\text{g of sample}} \times 100\% \qquad (23)$$

B. PERCENT (WEIGHT/VOLUME), or % (w/v)

This is the grams of pure analyte in 100 mL of solution. For example, a 10% (w/v) sucrose solution contains 10 g of sucrose per 100 mL. **The % (w/v) is calculated using equation 24, and the grams of pure analyte needed to prepare a solution with a given volume (mL) and percent concentration are calculated using equation 25, below:**

$$\% \text{ (w/v)} = \times \frac{\text{g of analyte in sample}}{\text{mL of sample}} \times 100\% \qquad (24)$$

$$\therefore \text{ g of analyte} = \frac{\% \text{ (w/v)} \times \text{mL of sample}}{100\%} \qquad (25)$$

Examples for Equations 23–25: See problems 13 and 15 in section 1.6.

C. MILLIGRAM PERCENT (mg %)

This is the milligrams of pure analyte in 100 mL of solution. For example, a 20 mg % solution of NaCl contains 20 mg of NaCl in 100 mL of the solution. **The mg % is calculated using equation 26, and the milligrams of pure analyte needed to prepare a solution with a given volume (mL) and mg % concentration is calculated using equation 27, below:**

$$\text{mg } \% = \frac{\text{mg of analyte in sample}}{\text{mL of sample}} \times 100\% \qquad (26)$$

---

[1]For the units in these equations to cancel out, the %, ppm, or ppb units in the equations must be substituted with their gram and/or volume equivalents. For example, when dealing with % (w/w), replace (% analyte) with (g of analyte/100 g of sample), and the % to the right of the equation with (g of sample/100 g of sample); and when dealing with % (w/v), replace (% analyte) with (g of analyte/100 mL of sample), and the % to the right with (mL of sample/100 mL of sample). See problems 14 and 15 in section 1.6.

$$\therefore \text{ mg of analyte } = \frac{\text{mg \% } \times \text{ mL of sample}}{100\%} \quad (27)$$

D. PARTS PER MILLION (WEIGHT/WEIGHT), or ppm (w/w)

This is the grams of pure analyte in 10^6 g of sample. For example, if trace Na^+ in solid $MgCl_2$ is stated as 10 ppm, then 10 g of Na^+ is contained in every 10^6 g of $MgCl_2$. **The ppm (w/w) is calculated using equation 28, below:**

$$\text{ppm (w/w)} = \frac{\text{g of analyte in sample}}{\text{g of sample}} \times 10^6 \text{ ppm} \quad (28)$$

Examples: See problems 16 and 17 in section 1.6.

E. PARTS PER MILLION (WEIGHT/VOLUME), or ppm (w/v)

This is the grams of pure analyte in 10^6 mL of sample. If, for example, the concentration of Na^+ in a standard solution is stated as 1000 ppm, the solution contains 1000 g of $Na^+/10^6$ mL. **The ppm (w/v) is calculated using equation 29, and the grams of pure analyte needed to prepare a solution with a given volume (mL) and ppm (w/v) concentration is calculated using equation 30, below:**

$$\text{ppm (w/v)} = \frac{\text{g of analyte in sample}}{\text{mL of sample}} \times 10^6 \text{ ppm} \quad (29)$$

$$\therefore \text{ g of analyte } = \frac{\text{ppm (w/v)} \times \text{ mL of sample}}{10^6 \text{ ppm}} \quad (30)$$

Examples: See problems 16 and 17 in section 1.6.

F. PARTS PER BILLION (WEIGHT/WEIGHT), or ppb (w/w)

This is the grams of pure analyte in 10^9 g of sample. Thus, if the concentration of trace Pb^{2+} in solid $CaCl_2$ is stated as 50 ppb, then 10^9 g of the $CaCl_2$ contains 50 g of Pb^{2+}. **The ppb (w/w) is calculated using equation 31, below:**

$$\text{ppb (w/w)} = \frac{\text{g of analyte in sample}}{\text{g of sample}} \times 10^9 \text{ ppb} \quad (31)$$

G. PARTS PER BILLION (WEIGHT/VOLUME), or ppb (w/v)

This is the grams of pure analyte in 10^9 mL of sample. Thus, if the concentration of trace K^+ in concentrated H_2SO_4 is stated as 100 ppb, the solution contains 100 g of $K^+/10^9$ mL. **The ppb (w/v) is calculated using equation 32, and the grams of pure analyte needed to prepare a solution with a given mL and ppb (w/v) concentration is calculated, below, using equations 33 and 34, respectively:**

$$\text{ppb (w/v)} = \frac{\text{g of analyte in sample}}{\text{mL of sample}} \times 10^9 \text{ ppb} \tag{32}$$

$$\therefore \text{ g of analyte} = \frac{\text{ppb (w/v)} \times \text{mL of sample}}{10^9 \text{ ppb}} \tag{33}$$

H. DENSITY (ρ) AND SPECIFIC GRAVITY (sp gr)

Density is the quantity of mass per unit volume of a substance. It is calculated using equation 34, with weight used to approximate mass.

$$\rho = \frac{\text{wt}}{\text{vol}} \tag{34}$$

Specific gravity is the density of a fluid relative to that of H_2O:

$$\text{sp gr} = \frac{\rho_{\text{sample}}}{\rho_{H_2O}} \tag{35}$$

Substituting 1 g/mL for the density of H_2O at 4°C into equation 36,

$$\text{sp gr} = \frac{\rho_{\text{sample}} \text{ (g/mL)}}{1 \text{ (g/mL)}} \tag{36}$$

Therefore, the specific gravity of a fluid is numerically equal to its density.

Examples: See problems 12 and 20 in section 1.6.

I. SPECIFIC VOLUME (sp vol)

This is the volume occupied by a unit weight of a solute when dissolved. It is calculated using equation 37, below:

$$\text{sp vol} = \frac{\text{vol of dissolved solute}}{\text{wt of solute}} = \frac{1}{\rho} \tag{37}$$

Therefore, the specific volume of a solute is equal to the inverse of its density.

1.4. Dilution Factor in Concentration and Volume Calculations

Often, one seeks the volume of a stock reagent that should be added to a known volume of diluent or assay medium to yield a desired concentration, or the final concentration of the reagent after adding a known volume to the diluent or assay medium. An equation relating the various parameters is derived as follows: Let V_1 and C_1, respectively, be the needed volume and the needed concentration of the stock reagent; and V_2, C_2, the final volume and desired the needed concentration, respectively, of the diluted reagent. Then,

$$\text{dilution factor} = \frac{C_1}{C_2} = \frac{V_2}{V_1} \tag{38}$$

$$\therefore\ C_1V_1 = C_2V_2 \tag{39}$$

or

$$V_1 = \frac{C_2V_2}{C_1}$$

Any of the parameters in equation 39 may be calculated, provided that the other three are known. In the calculations, C_1 and C_2 must have identical units; likewise, V_1 and V_2 must have identical units.

Examples: See problems 10d, 12, 20c, 22, and 23c in section 1.6.

1.5. Preparing Reagents

The steps involved in reagent preparation are outlined below and include references to relevant equations and examples in the text.

(1) Examine the purity, formula weight, and mole ratio data for the stock reagent

If the percent impurity of the stock reagent is high, apply a correction factor in step 2, below. This ensures that the concentration of the pure solute in the final solution is not significantly lower than the expected value (see problems 12 and 14 in section 1.6). If you have to calculate the formula weight, make sure that it contains the atomic weight of all the elements in the chemical formula (including bound H_2O and ions; see problem 11). When preparing a solution of a substance that is part of a compound, and 1 mole of the compound contains more than 1 mole of the substance, determine what molarity of the compound will yield the desired molarity of the substance. Use this value in step 2, below. For example, when preparing a 100 mM solution of OAc^- using $Mg(OAc)_2$, base the calculations on 50 mM $Mg(OAc)_2$, because 1 mole of $Mg(OAc)_2$ contains 2 mole of OAc^-. See also problem 8.

(2) Calculate the amount of the stock reagent needed

(a) Solid Reagents

If the stock reagent is a solid, calculate the amount needed according to the information provided in Table 1.1, below.

Table 1.1. Equations Used to Calculate Amount of Solid Reagent Needed to Prepare a Given Solution

Unit for Concn	Equation No.	Examples of Problems
mol/L (**M**)	9	7, 8, 11
mol/L (corrected for % impurity)	10	12, 14
mol/kg (**Molatity**)	11	9
equiv/L (**N**)	12	18, 20c, 22
% (w/v)	15	15
mg %	16	—
ppm (w/v)	17	—
ppb (w/v)	18	—
M or **N** (acid solutions)[a]	10, 12, 35	20–22
g/mL	g needed = C (g/mL) × V (mL)[b]	

[a] See step 5
[b] C and V denote concentration and volume, respectively.

(b) Liquid reagents

If the stock reagent is a liquid, calculate the volume needed, using equation 39. Whenever the unit of concentration for the reagent being prepared differs from that of the stock reagent, convert the latter to the former before applying equation 39. **Examples** are given in problems 10d, 12, 20c, 22, and 23c in section 1.6.

(3) Dissolve the stock reagent

Measure out and dissolve the calculated amount in a diluent volume that is about 80% of the target final volume of the solution[2]. Use a volumetric flask if available.

(4) Adjust the reagent volume

Add additional diluent to increase the volume to the target value.

(5) Caution

To protect against accidental chemical injury, dissolve concentrated acid and base solutions under a hood. Add the acid slowly to the H_2O and stir. Pouring H_2O rapidly into a concentrated acid solution causes a violent reaction that may lead to an explosion! Wear goggles to protect your eyes.

[2] If the diluent volume is equal to the target final volume prior to dissolving the stock reagent, the final volume of the solution will exceed the target final volume, resulting in a lower-than-expected concentration. This is because the solute molecules occupy a volume in the solution. See section 1.3.I.

1.6. Practical Examples

PROBLEM 1

(a) If there are 8.0 mg of Na^+ contaminant per liter of H_2O sample, calculate the number of moles of Na^+ per liter. The at. wt of Na is 23. (b) Calculate the weight of Na^+ in another sample of H_2O that contains 100 μmol Na^+/L.

SOLUTION:

(a) From equation 1,

$$\text{mol} = \frac{\text{wt (g)}}{\text{mol wt (g/mol)}}$$

$$\therefore \ \text{mol Na}^+/\text{L} = \frac{(0.0080 \text{ g Na}^+)/\text{L}}{23 \text{ g/mol}} = 3.5 \times 10^{-4} \text{ mol/L}$$

(b) From equation 1,

$$\text{mol} = \frac{\text{wt (g)}}{\text{mol wt (g/mol)}}$$

$$\therefore \ \text{wt of Na}^+/\text{L of H}_2\text{O} = \text{mol} \times \text{mol wt}$$

$$= 1 \times 10^{-4} \text{ mol} \times 23 \text{ g/mol}$$

$$= 2.3 \times 10^{-3} \text{ g}$$

$$100 \ \mu\text{mol} = 100 \ \mu\text{mol} \times (1 \text{ mol})/(1 \times 10^6 \ \mu\text{mol})$$

$$= 1 \times 10^{-4} \text{mol (see Appendix C, Table C-6)}$$

PROBLEM 2

If there are 8.0 mg of dissolved O_2 per liter of H_2O at 20°C and 1 atm, calculate (a) the number of moles of molecular O_2 and (b) the molarity of dissolved O_2 per liter of H_2O. The mol wt of O_2 is 32.

SOLUTION:

(a) Using equation 1,

$$\text{mol} = \frac{\text{wt (g)}}{\text{mol wt (g/mol)}}$$

$$\therefore \ \text{mol O}_2/\text{L} = \frac{(0.0080 \text{ g O}_2)/\text{L}}{32 \text{ g/mol}} = 2.5 \times 10^{-4} \text{ mol/L}$$

(b) Using equation 7,

$$\text{M of dissolved } O_2 = \frac{\text{mol}}{\text{L}} = \frac{2.5 \times 10^{-4}\ \text{mol}}{1\ \text{L}}$$

$$= 2.5 \times 10^{-4}\ \text{M or } 250\ \mu\text{M}$$

PROBLEM 3

Calculate the number of potassium (K) atoms in a 1.00 μg sample of pure KCl. The at. wt of K is 39.1 and the mol wt of KCl is 74.6.

SOLUTION:

$$\text{Ratio of K in 1 mol of KCl} = \frac{\text{at. wt of K}}{\text{mol wt of KCl}} = \frac{39.1}{74.6}$$

$$\therefore\ \text{g of K in 1.00 } \mu\text{g KCl} = \frac{39.1\ \text{g of K}}{74.6\ \text{g of KCl}} \times 1.00 \times 10^{-6}\ \text{g of KCl}$$

$$= 5.24 \times 10^{-7}\ \text{g of K}$$

Number of atoms in 1 mol of K (39.1 g) $= 6.022 \times 10^{23}$ atoms

$$\therefore\ \text{number of atoms in } 5.24 \times 10^{-7}\ \text{g of K}$$

$$= \text{mol of K} \times 6.022 \times 10^{23}\ \text{atoms/mol}$$

$$= \frac{5.24 \times 10^{-7}\ \text{g}}{39.1\ \text{g}} \times \frac{6.022 \times 10^{23}\ \text{atoms}}{1\ \text{mol}}$$

$$= 8.07 \times 10^{15}\ \text{atoms}$$

PROBLEM 4

Calculate the number of molecules of ampicillin per microliter of a 10.0 nM solution.

SOLUTION:

1 L or 1000 mL of the solution contains 10.0 nmol of ampicillin. Therefore, 1 μL of the solution contains

$$\frac{10.0 \text{ nmol}}{1000 \text{ mL}} \times (1 \times 10^{-3} \text{ mL}) = \frac{1.0 \times 10^{-8} \text{ mol}}{10^{3} \text{ mL}} \times (1 \times 10^{-3} \text{ mL})$$

$$= 1.0 \times 10^{-14} \text{ mol}$$

Number of molecules in 1 mol of ampicillin $= 6.022 \times 10^{23}$ molecules.

$$\therefore \text{ number of molecules in } 1 \times 10^{-14} \text{ mol ampicillin}$$

$$= \text{mol ampicillin} \times 6.022 \times 10^{23} \text{ molecules/mol}$$

$$= \frac{6.022 \times 10^{23} \text{ molecules}}{1 \text{ mol}} \times (1.0 \times 10^{-14} \text{ mol/μL})$$

$$= 6.0 \times 10^{9} \text{ molecules/μL}$$

PROBLEM 5

For 6.7 g of sodium acetate (NaOAc) and 6.7 g of $Mg(OAc)_2$, calculate (a) the moles of each salt, (b) the moles of acetate (OAc^-) in the NaOAc, and (c) the moles of OAc^- in the $Mg(OAc)_2$. The formula weights are: NaOAc, 82.03; $Mg(OAc)_2{\cdot}4H_2O$, 214.46.

SOLUTION:

(a) From equation 1,

$$\text{mol} = \frac{\text{wt (g)}}{\text{mol wt (g/mol)}}$$

$$\therefore \text{ mol NaOAc} = \frac{6.7 \text{ g}}{82.03 \text{ g/mol}} = 0.082 \text{ mol}$$

and

$$\text{mol } Mg(OAc)_2 = \frac{6.7 \text{ g}}{214.46 \text{ g/mol}} = 0.031 \text{ mol}$$

(b) Moles of OAc^- in each salt can be calculated in two ways:

(i) 1 mol of NaOAc contains 1 mol of OAc^-

$\therefore$ mol of OAc^- in 0.082 mol of NaOAc = 1 × 0.082 mol

= 0.082 mol

$$\text{(ii) mol of OAc}^-\text{/mol of NaOAc} = \frac{59.03 \text{ (mol wt of Ac)}}{82.03 \text{ (mol wt of NaOAc)}}$$

$$\therefore \text{ g OAc}^- \text{ in 6.7 g NaOAc} = \frac{59.03}{82.03} \times 6.7 \text{ g} = 4.82 \text{ g}$$

Using equation 1,

$$\text{mol OAc}^- \text{ in 6.7 g NaOAc} = \frac{\text{g of OAc}^-}{\text{mol wt of OAc}^-} = \frac{4.82 \text{ g}}{59.03 \text{ g/mol}}$$

= 0.082 mol

(c) Following the above examples (5b),

(i) 1 mol of $Mg(OAc)_2$ contains 2 mol of OAc^-

$\therefore$ mol of OAc^- in 0.031 mol of $Mg(OAc)_2$ = 2 × 0.031 mol

= 0.062 mol

(ii) As an exercise, calculate the moles of OAc^- in $Mg(OAc)_2$ using method (ii), above (answer = 0.062).

PROBLEM 6

If 6.7 g of sodium acetate (NaOAc) or 6.7 g of $Mg(OAc)_2$ is dissolved in a total volume of 0.200 L of H_2O, calculate (a) the molarity of NaOAc, (b) the molarity of $Mg(OAc)_2$, and (c) the molarity of OAc^- in each solution. Use the formula weights given in problem 5.

SOLUTION:

The problems can be solved using either equation 7 or 8.
(a) Using equation 8,

$$M = \frac{\text{wt}}{\text{mol wt} \times \text{L}}$$

$$\therefore \text{ M of NaOAc} = \frac{6.7 \text{ g}}{82.03 \text{ g/mol} \times 0.200 \text{ L}} = 0.41 \text{ mol/L}$$

(b) Following the example in (a),

$$\text{M of Mg(OAc)}_2 = \frac{6.7 \text{ g}}{214.46 \text{ g/mol} \times 0.200 \text{ L}} = 0.16 \text{ mol/L}$$

(c) There is 1 mole of OAc^-/mole of NaOAc.

$$\therefore \text{ M of OAc}^- \text{ in NaOAc} = 1 \times \text{M of NaOAc}$$

$$= 1 \times 0.41 \text{ mol/L} = 0.41 \text{ M}$$

There are 2 mole of OAc^-/mole of $Mg(OAc)_2$.

$$\therefore \text{ M of OAc}^- \text{ in Mg(OAc)}_2 = 2 \times \text{M of Mg(OAc)}_2$$

$$= 2 \times 0.16 \text{ mol/L} = 0.32 \text{ M}$$

PROBLEM 7

How many grams of $EDTA{\cdot}4H_2O$ (fw, 380) are needed to prepare 0.100 L of a 0.020 M solution.

SOLUTION:

Using equation 9,

$$\text{g of EDTA needed} = \text{M} \times \text{mol wt} \times \text{L}$$

$$= 0.020 \text{ mol/L} \times 380 \text{ g/mol} \times 0.100 \text{ L}$$

$$= 0.76 \text{ g}$$

PROBLEM 8

Prepare 0.100 L of a 0.050 M solution of OAc^- using (a) NaOAc (fw, 82.03), or (b) $Mg(OAc)_2$ (fw, 214.46).

SOLUTION:

(a) There is 1 mole of OAc^-/mole of NaOAc. Therefore, 0.050 M OAc^- is equivalent to 0.050 M NaOAc. From equation 9,

$$\text{wt} = \text{M} \times \text{mol wt} \times \text{L}$$

$$\therefore \text{ g of NaOAc needed} = 0.050 \text{ mol/L} \times 82.03 \text{ g/mol} \times 0.100 \text{ L}$$

$$= 0.41 \text{ g}$$

(b) There are 2 mole of OAc^-/mole of $Mg(OAc)_2$. Therefore, 0.050 M OAc^- is equivalent to 50/2 or 0.025 M $Mg(OAc)_2$. Using equation 9,

$$\text{g of } Mg(OAc)_2 \text{ needed} = 0.025 \text{ mol/L} \times 214.46 \text{ g/mol} \times 0.100 \text{ L}$$

$$= 0.54 \text{ g}$$

Each salt is then dissolved in about 90 mL of H_2O, after which the volume is increased to 100 mL with additional H_2O.

PROBLEM 9

Prepare 500 mL of a 2.0 molal aqueous solution of KCl.

SOLUTION:

Using equation 13, calculate the weight of KCl needed:

$$\text{wt (g)} = \text{molality} \times \text{mol wt} \times \text{kg of solvent}$$

The mol wt of KCl is 74.55 g/mol, and 500 mL of H_2O weighs 0.500 kg.

$$\therefore \text{ wt of KCl needed} = 2.0 \text{ mol/kg} \times 74.55 \text{ g/mol} \times 0.500 \text{ kg}$$

$$= 74.5 \text{ g}$$

To prepare the solution, dissolve 74.5 g of KCl in 400 mL of H_2O, and then adjust the volume to 500 mL with additional H_2O.

PROBLEM 10

The concentration of ampicillin in a stock solution is 5.00×10^3 µg/mL. Calculate: (a) µg of ampicillin in 0.100 mL of the solution, (b) µmol of ampicillin in 0.100 mL of the solution, (c) molarity of ampicillin in the solution, and (d) prepare 30.0 mL of a working solution containing 10.0 µg/mL. The fw of the ampicillin is 349.41.

SOLUTION:

(a) 1 mL of the solution contains 5.00×10^3 µg

$$\therefore \text{ 0.100 mL contains } \frac{5.00 \times 10^3 \ \mu\text{g}}{1 \text{ mL}} \times 0.100 \text{ mL}$$

$$= 5.00 \times 10^2 \ \mu\text{g or } 5.00 \times 10^{-4} \text{ g of ampicillin}$$

(b) Using the result in (a), and equation 1,

$$\text{mol of ampicillin} = \frac{\text{g}}{\text{g-mol wt}} = \frac{5.00 \times 10^{-4}\ \text{g}}{349.41\ \text{g/mol}}$$

$$= 1.43 \times 10^{-6}\ \text{mol or } 1.43\ \mu\text{mol}$$

(c) From the result in (b), 0.100 mL (or 1.00×10^{-4} L) of the solution contains 1.43×10^{-6} mol. Then, using equation 7,

$$\text{M of ampicillin} = \frac{\text{mol}}{\text{L}} = \frac{1.43 \times 10^{-6}\ \text{mol}}{1.00 \times 10^{-4}\ \text{L}}$$

$$= 0.0143\ \text{M} \quad \text{or} \quad 14.3\ \text{mM}$$

The molarity can also be calculated using the result from (a) and equation 10:

$$\text{M} = \frac{\text{g of ampicillin}}{\text{mol wt} \times \text{L}} = \frac{5.00 \times 10^{-4}\ \text{g}}{349.41\ \text{g/mol} \times 10^{-4}\ \text{L}} = 0.0143\ \text{M}$$

(d) Let V_1 be the volume of the stock ampicillin solution needed to prepare the 30.0 mL (V_2) solution, and let C_1 and C_2 be the concentrations of ampicillin in the stock solution and the 30.0 mL solution, respectively. Then, using equation 39,

$$\therefore\ V_1 = \frac{C_2 V_2}{C_1} = \frac{10.0\ \mu\text{g/mL} \times 30.0\ \text{mL}}{5.00 \times 10^{3}\ \mu\text{g/mL}}$$

$$= 0.0600\ \text{mL} \quad \text{or} \quad 60.0\ \mu\text{L}$$

To prepare the solution, add 60 µL of the ampicillin solution to 29.94 mL of diluent.

PROBLEM 11

How many grams of ATP are needed to prepare 0.200 L of a 100 µM solution? The anhydrous mole wt of ATP is 487.2; the ATP powder being used has 3 moles of H_2O and 2 moles of Na^+ per mole.

SOLUTION:

First calculate the formula weight of the ATP:

$$\text{fw} = \text{mol wt of ATP} + 3 \text{ mol wt of } H_2O + 2 \text{ mol wt of } Na^+$$

$$= 487.2 + (3 \times 18) + (2 \times 23) = 587.2$$

Then, calculate the grams of ATP needed, using equation 9:

$$\text{g of ATP} = \text{M} \times \text{mol wt} \times \text{L}$$

$$= 0.0001 \text{ mol/L} \times 587.2 \text{ g/mol} \times 0.200 \text{ L}$$

$$= 0.012 \text{ g}$$

PROBLEM 12

ß-Mercaptoethanol is available as a 98% (w/w) solution (sp gr 1.114). What volume is needed to prepare 0.100 L of a 50 mM solution? Assume an fw of 78.13.

SOLUTION:

The volume can be calculated in two ways:

(i) Calculate the grams of ß-mercaptoethanol needed and how many milliliters of the stock solution are equivalent to the calculated weight. Using equation 10,

$$\text{g of ß-mercaptoethanol needed} = \text{M} \times \text{mol wt} \times \text{L} \times \frac{100\ \%}{\%\ \text{purity}}$$

$$= \frac{0.05 \text{ mol/L} \times 78.13 \text{ g/mol} \times 0.100 \text{ L} \times 100\%}{98\%} = 0.399 \text{ g}$$

Using equation 34,

$$\text{vol} = \frac{\text{wt}}{\rho} = \frac{0.399 \text{ g}}{1.114 \text{ g/mL}} = 0.358 \text{ mL}$$

(ii) Calculate the molarity of ß-mercaptoethanol in the stock solution. Use it to calculate how many milliliters should be diluted to 100 mL to yield 50 mM.

The 98% solution contains 98 g of ß-mercaptoethanol per 100 g of solution. Using equation 34,

$$\text{vol of 100 g solution} = \frac{\text{wt}}{\rho} = \frac{100 \text{ g}}{1.114 \text{ g/mL}} = 89.77 \text{ mL}$$

From equation 8,

$$M = \frac{\text{wt}}{\text{mol wt} \times \text{L}} = \frac{98\ \text{g}}{78.13\ \text{g/mol} \times 0.08977\ \text{L}} = 13.97\ \text{M}$$

Let V_1 be the volume of the stock ß-mercaptoethanol needed to prepare the 100 mL (V_2) solution, and let C_1 and C_2 be the concentrations of ß-mercaptoethanol in the stock solution and the 100 mL solution, respectively. Then, using equation 39,

$$V_1 = \frac{C_2V_2}{C_1} = \frac{0.05\ \text{M} \times 0.100\ \text{L}}{13.97\ \text{M}} = 0.358\ \text{mL}$$

PROBLEM 13

One g of dry peptone was analyzed and found to contain 50.0 μg of K^+. Calculate (a) the % (w/w) of K^+ in the peptone, (b) the % (w/v) of K^+ in a 5% (w/v) solution of the peptone, and (c) the molarity of K^+ in the 5% solution. The at. wt of K is 39.1.

SOLUTION:

(a) From equation 23,

$$\%\ (\text{w/w}) = \frac{\text{g of analyte in sample}}{\text{g of sample}} \times 100\%$$

$$\therefore\ \%\ (\text{w/w})\ \text{of K} = \frac{5.00 \times 10^{-5}\ \text{g of K}}{1.0\ \text{g of peptone}} \times 100\% = 0.0050\%$$

(b) There are 5 g of peptone per 100 mL of a 5% solution and 50.0 μg of K per gram of peptone.

$$\therefore\ \mu\text{g of K in 5 g of peptone} = \frac{50.0\ \mu\text{g of K}}{1\ \text{g of peptone}} \times 5\ \text{g of peptone}$$

$$= 2.50 \times 10^{2}\ \mu\text{g of K}$$

Using equation 24,

$$\%\ (\text{w/v})\ \text{of K} = \frac{\text{g of analyte}}{100\ \text{mL of sample}} \times 100\%$$

$$= \frac{2.50 \times 10^{-4}\ \text{g of K}}{100\ \text{mL (pep.)}} \times 100\ \% = 0.000250\%.$$

(c) From the result in (b), the 5% solution contains 2.50×10^{-4} g of K per 100 mL (0.100 L). Using equation 8,

$$M = \frac{\text{wt}}{\text{mol wt} \times \text{L}} = \frac{2.50 \times 10^{-4}\ \text{g}}{39.1\ \text{g/mol} \times 0.100\ \text{L}} = 6.39 \times 10^{-5}\ \text{M}$$

PROBLEM 14

A tetraphenyl phosphonium chloride (TPPCl) powder (fw, 342.39) is 96% pure. How many grams are needed to prepare 0.100 L of a 10.0 mM solution?

SOLUTION:

Using equation 10,

g of TPPCl needed

$$= \text{M} \times \text{mol wt} \times \text{L} \times \frac{100\%}{\%\ \text{purity}}$$

$$= \frac{0.010\ \text{mol/L} \times 342.39\ \text{g/mol} \times 0.100\ \text{L} \times 100\ \%}{96\ \%}$$

$$= 0.357\ \text{g}$$

PROBLEM 15

How many grams of *n*-octyl glucoside are needed to prepare a 5.0% (w/v) solution if the octyl glucoside is (a) 100% (w/w) pure and (b) 95% (w/w) pure?

SOLUTION:

(a) When 100% purity is assumed, a 5.0% (w/v) solution contains 5.0 g in 100 mL of the solution.

∴ 5.0 g of the 100% pure *n*-octyl glucoside is needed to prepare 100 mL of the 5.0% solution.

Alternatively, the problem can be solved by using equation 23:[3]

$$\%\ (\text{w/w}) = \frac{\text{g of analyte}}{\text{g of sample}} \times 100\%$$

---

[3] See footnote 1.

$$\therefore \text{ g of sample} = \frac{5.0 \text{ g of analyte} \times 100}{100}$$

$$= 5.0 \text{ g sample}$$

To prepare 100 mL of a 5.0% (w/v) solution, 5.0 g of the 100% pure *n*-octyl glucoside is needed.

(b) The amount of the 95% (w/w) sample containing 5.0 g of pure *n*-octyl glucoside is similarly calculated using equation 23:[4]

$$\% \text{ (w/w)} = \frac{\text{g of analyte}}{\text{g of sample}} \times 100\%$$

$$\therefore \text{ g of sample} = \frac{5.0 \text{ g of analyte} \times 100}{95}$$

$$= 5.3 \text{ g of sample}$$

To prepare 100 mL of a 5% (w/v) solution, 5.3 g of the 95% *n*-octyl glucoside is needed.

PROBLEM 16

The concentration of Na^+ in a standard solution is 1000 ppm (w/v). What is the molarity of Na^+? The at. wt of Na is 23.

SOLUTION:

First, calculate the grams of Na^+ per liter of solution.
By rearranging equation 29,[5]

$$\text{g of analyte} = \frac{\text{ppm of analyte (w/v)} \times \text{mL of sample}}{10^6 \text{ ppm}}$$

$$\therefore \text{ g of Na}^+\text{/L} = \frac{1000 \text{ ppm} \times 1000 \text{ mL of solution}}{10^6 \text{ ppm}}$$

$$= \frac{1000 \times 1000 \text{ mL of solution}}{10^6}$$

$$= 1 \text{ g of Na}^+$$

---

[4] See footnote 1.

[5] See footnote 1.

Second, calculate the M of Na^+, using equation 8:

$$\text{M of Na}^+ = \frac{\text{wt}}{\text{at. wt} \times \text{L}} = \frac{1 \text{ g}}{23 \text{ g/mol} \times 1 \text{ L}} = 0.04 \text{ M}$$

The standard solution is 0.04 M in Na^+.

PROBLEM 17

Trace Pb^{2+} in $MgSO_4$ is 500 ppm (w/w). Calculate the molarity of Pb^{2+} in a 20% (w/v) solution of the $MgSO_4$. The at. wt of Pb is 207.2.

SOLUTION:

First, calculate the grams of Pb^{2+} in the 20 g of $MgSO_4$, needed to prepare a 20% (w/v) solution. From equation 23 (rearranged),[6]

$$\text{g of analyte} = \frac{\text{ppm (w/w) of analyte} \times \text{g of sample}}{10^6 \text{ ppm of sample}}$$

$$\therefore \text{ g of Pb}^{2+}/20 \text{ g of MgSO}_4$$

$$= \frac{500 \text{ ppm of Pb}^{2+} \times 20 \text{ g of Pb}^{2+}}{10^6 \text{ ppm}} = 0.010 \text{ g}$$

Second, using equation 8, calculate the molarity of Pb^{2+}.

$$\text{M of Pb}^{2+} = \frac{\text{wt}}{\text{mol wt} \times \text{L}} = \frac{0.010 \text{ g}}{207.2 \text{ g/mol} \times 0.100 \text{ L}}$$

$$= 5 \times 10^{-4} \text{ M} \quad \text{or} \quad 500 \text{ }\mu\text{M}$$

PROBLEM 18

Prepare 0.500 L of a 0.200 N solution of (a) NaOH and (b) $Ca(OH)_2$. The fw is 40.0 for NaOH, and 74.1 for $Ca(OH)_2$.

SOLUTION:

(a) NaOH contains 1 OH^-/molecule. Therefore, the number of equiv/mole $= n = 1$. Using equation 3,

$$\text{equiv wt of NaOH} = \frac{\text{mol wt}}{n} = \frac{40.0 \text{ g/mol}}{1 \text{ equiv/mol}} = 40.0 \text{ g/equiv}$$

---

[6] See footnote 1.

Using equation 18,

$$\text{g of NaOH needed} = \text{N} \times \text{equiv wt} \times \text{L}$$

$$= 0.200 \text{ equiv/L} \times 40.0 \text{ g/equiv} \times 0.500 \text{ L}$$

$$= 4.00 \text{ g}$$

Dissolve 4.00 g of NaOH in 450 mL of H_2O. Adjust volume to 500 mL.

(b) Similarly, $Ca(OH)_2$ contains 2 OH^-/molecule.

$$\therefore\ n = 2$$

$$\text{equiv wt of } Ca(OH)_2 = \frac{74.1 \text{ g/mol}}{2 \text{ equiv/mol}} = 37.0 \text{ g/equiv}$$

$$\text{g } Ca(OH)_2 \text{ needed} = 0.200 \text{ equiv/L} \times 37.0 \text{ g/equiv} \times 0.500 \text{ L}$$

$$= 3.70 \text{ g}$$

Dissolve 3.70 g of $Ca(OH)_2$ in 450 mL of H_2O. Adjust volume to 500 mL.

PROBLEM 19

Calculate the number of (a) base equivalents in 5.0 g of NaOH, (b) acid equivalents in 1.0 mL of a 0.10 N H_2SO_4 solution, and (c) electron equivalents in 50.0 mL of a 10.0 mM solution of a compound containing 1 mole of transferable electrons per mole. The mole wt of NaOH is 40.0.

SOLUTION:

(a) NaOH contains 1 base equiv/mole (i.e., 1 OH^-/molecule); $n = 1$. Using equation 3,

$$\text{equiv wt of NaOH} = \frac{\text{mol wt}}{n} = \frac{40.0 \text{ g/mol}}{1 \text{ equiv/mol}} = 40.0 \text{ g/equiv}$$

Using equation 4,

$$\text{No. of equiv} = \frac{\text{wt}}{\text{equiv wt}} = \frac{5.0 \text{ g}}{40.0 \text{ g/equiv}} = 0.12 \text{ equiv}$$

(b) A 0.10 N solution contains 0.10 equiv/L; i.e., 1000 mL of 0.10 N H_2SO_4 contains 0.10 equiv

$$\therefore \ 1.0 \text{ mL contains: } \frac{0.10 \text{ equiv}}{1000 \text{ mL}} \times 1.0 \text{ mL} = 1.0 \times 10^{-4} \text{ equiv}$$

(c) The compound contains 1 mole of transferable electrons/mol.

$$\therefore \text{ The number of equiv/mol} = n = 1; \text{ M} = 0.0100.$$

Using equation 21,

$$\text{N} = \text{nM} = 1 \text{ equiv/mol} \times 0.0100 \text{ mol/L} = 0.0100 \text{ equiv/L}$$

This means that 1000 mL contains 0.0100 electron equiv

$$\therefore \ 50.0 \text{ mL contains } \frac{0.0100 \text{ equiv}}{1000 \text{ mL}} \times 50.0 \text{ mL}$$

$$= 5.00 \times 10^{-4} \text{ equiv}$$

PROBLEM 20

The concentration of HCl in a commercial concentrated HCl solution is given as 37.0% (w/w). Calculate (a) the molarity and (b) the normality of HCl in the solution. (c) Prepare 100.0 mL of a 2.0 N solution of the HCl. The fw and sp gr of the HCl are 36.5 and 1.19, respectively.

SOLUTION:

(a) First, calculate the volume of 100.0 g of the HCl solution. Using equation 34,

$$\text{vol of 100 g of solution} = \frac{\text{g of solution}}{\rho \text{ of solution}} = \frac{100.0 \text{ g}}{1.19 \text{ g/mL}} = 84.03 \text{ mL}$$

Second, calculate the molarity using equation 8; 100 g of the solution contains 37.0 g of HCl in a volume of 0.0840 L.

$$\therefore \ \text{M of HCL} = \frac{\text{wt}}{\text{mol wt} \times \text{L}} = \frac{37.0 \text{ g}}{36.5 \text{ g/mol} \times 0.08403 \text{ L}} = 12.1 \text{ M}$$

(b) HCl contains 1 H^+ per molecule. The number of equiv/mol $= n = 1$.

Using equation 21,

$$N = nM$$

$$= 1 \text{ equiv/mol} \times 12.06 \text{ mol/L}$$

$$= 12.06 \text{ equiv/L or } 12.1 \text{ N}$$

(c) Calculate the volume of concentrated HCl needed, using equation 39: Let V_1 be the volume of concentrated HCl needed to prepare 100.0 mL (V_2) of the 2.0 N solution; C_1 and C_2 are the concentrations of HCl in the concentrated and 2 N solutions, respectively. From equation 39,

$$C_1V_1 = C_2V_2$$

$$\therefore V_1 = \frac{C_2V_2}{C_1} = \frac{2.0 \text{ N} \times 100.0 \text{ mL}}{12.06 \text{ N}} = 17 \text{ mL}$$

Add 17 mL of concentrated HCl to 83 mL of H_2O to prepare 100.0 mL of the 2.0 N solution.

PROBLEM 21

Calculate the normality of commercial concentrated H_2SO_4 and H_3PO_4. The % concentrations, fw, and sp gr, respectively, are: H_2SO_4, 95.0% (w/w), 98.1, 1.84; H_3PO_4, 85% (w/w), 98, 1.71.

SOLUTION:

Calculate the molarity of H_2SO_4 and H_3PO_4 following the example in problem 19a. These values are: H_2SO_4, 17.8 M; H_3PO_4, 14.7 M.

H_2SO_4 contains 2 H^+/molecule; the number of equiv/mol $= n = 2$. H_3PO_4 contains 3 ionizable H^+/molecule; the number of equiv/mol $= n = 3$. The normality is calculated using equation 21,

$$N = nM$$

For H_2SO_4,

$$N = 2 \text{ equiv/mol} \times 17.8 \text{ mol/L} = 35.6 \text{ equiv/L or } 35.6 \text{ N}$$

For H_3PO_4,

$$N = 3 \text{ equiv/mol} \times 14.7 \text{ mol/L} = 44.1 \text{ equiv/L or } 44.1 \text{ N}$$

PROBLEM 22

Prepare 100.0 mL of a 2.0 N solution of (a) H_2SO_4 and (b) H_3PO_4, starting with the concentrated acids.

SOLUTION:

Calculate the normality of concentrated H_2SO_4 and H_3PO_4 (problem 20) or obtain it from Appendix D.1. These values are 35.6 N for H_2SO_4 and 44.1 N for H_3PO_4.

(a) Let V_1 be the volume of concentrated H_2SO_4 needed to prepare 100.0 mL (V_2) of the 2.0 N solution; C_1 and C_2 are the concentrations of H_2SO_4 in the concentrated and 2.0 N solutions, respectively. From equation 39,

$$C_1V_1 = C_2V_2$$

$$V_1 = \frac{C_2V_2}{C_1} = \frac{2.0\text{ N} \times 100.0\text{ mL}}{35.6\text{ N}} = 5.6\text{ mL}$$

Add 5.6 mL of concentrated H_2SO_4 to 94.4 mL of H_2O to prepare 100.0 mL of the 2.0 N solution.

(b) As an exercise, calculate the volume of H_3PO_4 needed to prepare the 100.0 mL of the 2.0 N solution (answer = 4.5 mL).

PROBLEM 23

A 75.0 µL aliquot of a solution of EcoR1 (100.0 units/mL) is added to 0.50 mL of buffer. Calculate (a) the dilution factor (DF) and (b) the final concentration of EcoR1 in the buffer. (c) Prepare 1.50 mL of a 15.0 units/mL solution starting with the 100.0 units/mL stock.

SOLUTION:

(a) From equation 38,

$$\text{DF} = \frac{V_2}{V_1} = \frac{0.50\text{ mL} + 0.0750\text{ mL}}{0.0750\text{ mL}} = 7.7$$

(b) Let V_1 be the volume of the 100.0 units/mL (C_1) solution added to the 0.50 mL (V_2) buffer, and let C_2 be the final concentration of EcoR1. Then, using equation 39, the final concentration is:

$$C_2 = \frac{C_1 V_1}{V_2} = \frac{100.0\ (\text{units/mL}) \times 0.0750\ \text{mL}}{0.575\ \text{mL}}$$

$$= 13.0\ \text{units/mL}$$

Alternatively, the final concentration can be calculated as follows:

$$C_2 = \frac{C_1}{\text{DF}} = \frac{100.0\ \text{units/mL}}{7.7} = 13.0\ \text{units/mL}$$

(c) Let V_1 be the volume of the stock solution needed to prepare 1.50 mL (V_2) of the 15.0 units/mL solution. C_1 and C_2 are the concentrations of EcoR1 in the stock and the 15.0 units/mL solutions, respectively. From equation 39,

$$V_1 = \frac{C_2 V_2}{C_1} = \frac{15.0\ \text{units/mL} \times 1.50\ \text{mL}}{100.0\ \text{units/mL}} = 0.225\ \text{mL}$$

Add 225 µL of the stock solution to 1.275 mL of diluent to prepare 1.50 mL of the 15.0 units/mL solution.

Chapter 2

BUFFERS: PRINCIPLES, CALCULATIONS, AND PREPARATION

2.1. Principles

A. DEFINITIONS AND ABBREVIATIONS

The function of pH buffers is to neutralize small changes in H^+ and OH^- concentrations, thereby maintaining pH fairly constant. The basics of pH buffering and buffer calculations and preparation are discussed below. The necessary symbols and their meanings are as follows: A weak acid and its conjugate base are represented by HA and A^-, respectively; a weak base and its conjugate acid are represented by R-NH_2 and R-NH_3^+, respectively, because the most commonly used weak base buffers have NH_2 basic functional group(s). In aqueous solutions, H^+ is hydrated and, therefore, is written as H_3O^+ when necessary. Molar concentration is indicated by enclosing the substance of interest in square brackets; for example, $[H^+]$ represents the molar concentration of H^+.

B. PROPERTIES OF ACIDS AND BASES

For the purpose of this chapter, acids and bases are defined according to Bronsted-Lowry (*2, 3*). An acid is a chemical compound that donates a H^+, and a base is a compound that accepts a H^+. When an acid is dissolved in H_2O, it donates its H^+ to a H_2O molecule which, in this case, acts as a base. When a base is dissolved, it accepts a H^+ from a H_2O molecule which, in this case, acts as an acid. These reactions are shown below using the symbols HA and R-NH_2 for the acid and base, respectively:

$$HA + H_2O \rightleftharpoons H_3O^+ + A^-$$

$$R\text{-}NH_2 + H_2O \rightleftharpoons R\text{-}NH_3^+ + OH^-$$

In the first equation, the ionized acid (A^-) is termed the conjugate base of HA, and HA is the conjugate acid of A^-. In the second equation, the ionized base ($R\text{-}NH_3^+$) is the conjugate acid of $R\text{-}NH_2$, and $R\text{-}NH_2$ is the conjugate base of $R\text{-}NH_3^+$.

As indicated by the double arrows in the equations above, acid-base reactions are reversible, but the extent of reversibility depends on the strength of the acid or base. Strong acids or bases dissociate almost completely in H_2O. For example, when the strong acid HCl is dissolved in H_2O, almost 100% of it dissociates as follows:

$$HCl + H_2O \rightleftharpoons H_3O^+ + Cl^-$$

where the longer upper arrow indicates a strong preference toward formation of the H_3O^+ and Cl^-. In a dilute solution at equilibrium, very little of the HCl remains, and H_3O^+ and Cl^- are the predominant species. The Cl^-, which is the conjugate base of HCl, is a weak base. Similarly, when a strong base is dissolved, the resultant conjugate acid is typically a weak one. **In general, a strong acid produces a weak conjugate base, and a strong base produces a weak conjugate acid.** As explained in the next section, this tendency toward complete ionization, and the production of weak conjugate forms, makes strong acids and bases unsuitable for pH buffering. In contrast to the above properties, a weak acid or base dissociates partially in H_2O. As such, significant quantities of both the conjugate acid and the conjugate base remain at equilibrium. Furthermore, **a weak acid produces a strong conjugate base, and a weak base produces a strong conjugate acid.** These properties make them suitable for pH buffering, the mechanism of which is discussed below.

C. MECHANISM OF pH BUFFERING BY WEAK ACIDS AND BASES

As mentioned above, weak acids and weak bases ionize incompletely when dissolved in H_2O. The solution, therefore, contains significant concentrations of the conjugate acid HA or $R\text{-}NH_3^+$ in equilibrium with H_3O^+ and the conjugate base (A^- or $R\text{-}NH_2$), as shown in equations 40–44, below.

For a weak acid with one ionizable H,

$$HA + H_2O \rightleftharpoons H_3O^+ + A^- \qquad (40)$$

$$K = \frac{[H_3O^+][A^-]}{[H_2O][HA]} \qquad (41)$$

where K is the equilibrium constant. If the ionization of H_2O is ignored, equation 40 becomes

$$HA \rightleftharpoons H^+ + A^- \qquad (42)$$

For a weak base that accepts one H^+,

$$R\text{-}NH_2 + H_2O \rightleftharpoons R\text{-}NH_3^+ + OH^- \quad (43)$$

$$K = \frac{[R\text{-}NH_3^+][OH^-]}{[R\text{-}NH_2][H_2O]} \quad (44)$$

where K is is the equilibrium constant.

When a small quantity of H^+ is added to a solution of a weak acid or base, most of the H^+ released, reacts with the conjugate base (A^- or $R\text{-}NH_2$) to form the corresponding conjugate acid (HA or $R\text{-}NH_3^+$). If OH^- is added, most of it reacts with the conjugate acid to form the conjugate base and H_2O. Thus, most of the added H^+ or OH^- are not free to appreciably change the total H^+ concentration in the solution and, as such, there is little or no change in the pH of the solution. The weak acid or base is, therefore, a pH buffer since it guards against changes in the pH of the solution. The resistance to pH changes is greatest at the pH where the concentrations of the conjugate acid and conjugate base are equal. This pH is known as the pK_a of the buffer (see Figure 2.1, section 2.2.B, and Appendix A.2). The effective buffering range is about $pK_a \pm 1$ pH unit (Figure 2.1), and for this reason, a buffer chosen for a biochemical assay should have a pK_a near the optimum pH for the assay.

A strong acid or base (e.g., HCl or KOH) cannot buffer pH because it dissociates almost completely in dilute aqueous solutions, and the conjugate base or acid is weak. If, for example, a small amount of H^+ is added to a dilute solution of HCl, the total H^+ concentration will increase because the conjugate base Cl^- is too weak to react with the added H^+. If OH^- is added, it reacts with free H^+, and since the concentration of HCl in the solution is too low to ionize and replace the reacting H^+, the concentration of H^+ in the solution decreases. In either case, there is no pH buffering.

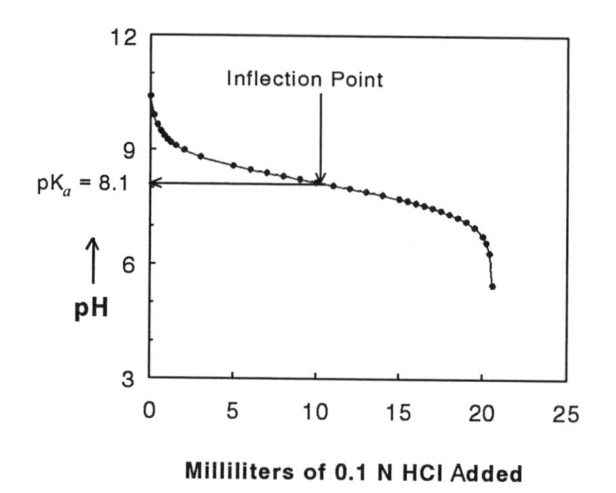

Figure 2.1. The pH titration curve for the monoprotic weak base, Tris(hydroxymethyl)aminomethane (Tris). Twenty-five mL of 0.1 M Tris was titrated with small volumes of 0.1 M HCl. The pH of the solution was plotted against the volume of HCl. At the inflection point of the curve, the concentrations of the conjugate base and the conjugate acid are equal, and pH buffering is maximal. This pH (8.1) is the pK_a of Tris.

2.2. Buffer, pH, and pOH Calculations

A. THE pH/pOH SCALE

The solvent for biological pH buffers is H_2O, and for this reason, the pH/pOH scale is based on the ionization constant of pure H_2O. Pure H_2O ionizes slightly as follows:

$$2H_2O \rightleftharpoons H_3O^+ + OH^- \tag{45}$$

or simply,

$$H_2O \rightleftharpoons H^+ + OH^- \tag{46}$$

The ionization constant, K_w (derived in Appendix A.1), is

$$K_w = [H^+][OH^-] \tag{47}$$

In pure H_2O at room temperature,

$$[H^+] = [OH^-] = 1 \times 10^{-7}\ M$$

$$\therefore\ [H^+][OH^-] = [1 \times 10^{-7}]^2 = 1 \times 10^{-14} \tag{48}$$

Equation 48 defines the range for the acidity or basicity of aqueous solutions. This range can vary over extreme values, from 1 M or greater to 1×10^{-14} M or less. Data with such a wide range of variations are difficult to handle on a linear scale and are therefore expressed as logarithms to compress the range and ease data handling. Thus, the concentrations of H^+ and OH^- in aqueous solutions are normally expressed as logarithms: pH and pOH, respectively. The scale for pH/pOH is derived by taking the logarithm of equation 48:

$$\log [H^+][OH^-] = \log [1 \text{ x } 10^{-14}]$$

or

$$\log [H^+] + \log [OH^-] = -14 \tag{49}$$

Multiplying both sides of equation 49 by –1 yields

$$-\log [H^+] + -\log [OH^-] = 14 \tag{50}$$

In general, –log X is defined as pX.

$$\therefore\ -\log [H^+] = \text{pH} \tag{51}$$

and

$$-\log [OH^-] = \text{pOH} \tag{52}$$

Substituting equations 51 and 52 into equation 50 yields

$$\text{pH} + \text{pOH} = 14 \tag{53}$$

which defines the pH/pOH scale for aqueous solutions. The negative sign in equations 51 and 52, which was arbitrarily introduced by multiplying equation 49 by –1, is intended to make the pH and the pOH values positive since most $[H^+]$ and $[OH^-]$ used in the laboratory are less than 1 M and would otherwise yield negative pH or pOH values. By this convention, a negative pH or pOH will result when $[H^+]$ or $[OH^-]$ is greater than 1 M.

Not all ions of a substance in solution are active; hence, the effective concentration or **activity** of an ion is less than the molar concentration. For this reason, pH and pOH are more accurately defined as

$$\text{pH} = -\log \gamma [H^+] \quad \text{and} \quad \text{pOH} = -\log [OH^-] \tag{54}$$

where γ is the activity coefficient of the respective ions. In dilute solutions (0.1 M or less), $\gamma \approx 1$. Therefore, $\gamma[H^+]$ and $\gamma[OH^-] \approx [H^+]$ and $[OH^-]$, respectively, and equations 51 and 52 may be used as written. A pH electrode measures the activity of H^+, i.e., $\gamma[H^+]$.

Equations 51–54 are used to calculate the pH or pOH of an aqueous solution provided that the *free* $[H^+]$ or $[OH^-]$ in the solution is known.

Example: See problem 27 and a related example, problem 28 in section 2.4.

B. THE HENDERSON-HASSELBACH EQUATION

The Henderson-Hasselbach equation, derived in Appendix A.2, is shown below. For a weak acid,

$$\text{pH} = \text{p}K_a + \log \frac{[\text{A}^-]}{[\text{HA}]} \tag{55}$$

For a weak base,

$$\text{pH} = \text{p}K_a + \log \frac{[\text{R-NH}_2]}{[\text{R-NH}_3^+]} \tag{56}$$

The equation relates the pH of a buffer to the pK_a and concentrations of the conjugate acid and conjugate base, and predicts that the pH of a buffer is determined by the (conjugate acid)/(conjugate base) ratio. In other words, a given pH for a buffer is attained once the (conjugate base)/(conjugate acid) ratio predicted for that pH by this equation is established. This implication underlies the methods by which pH buffers are prepared (see below). When $[A^-] = [HA]$, the ratio $[A^-]/[HA]$ becomes 1, log 1 = 0, and equation 55 becomes, pH = pK_a. This means that pK_a is the pH at which the concentrations of the conjugate acid and conjugate base are equal.

Buffer calculations often recuire calculating the $[A^-]/[HA]$ ratio. An expression for a direct calculation of this ratio is derived by solving equation 55 as follows:

$$\text{pH} = \text{p}K_a + \log \frac{[\text{A}^-]}{[\text{HA}]}$$

$$\log \frac{[\text{A}^-]}{[\text{HA}]} = \text{pH} - \text{p}K_a \tag{57}$$

Upon taking the antilogy,

$$\frac{[\text{A}^-]}{[\text{HA}]} = 10^{(\text{pH} - \text{p}K_a)} \tag{58}$$

For a weak base, equation 58 is similarly solved to yield:

$$= \frac{[\text{R-NH}_2]}{[\text{R-NH}_3^+]} \quad 10^{(\text{pH} - \text{p}K_a)} \tag{59}$$

Equations 55–59, in conjunction with equation 11, are used to calculate the amounts of the conjugate acid and conjugate base which, when dissolved, yield a buffer of a given pH, concentration, and volume.

Examples: See problems 24 and 25 in section 2.4.

C. BUFFER CAPACITY (BC)

Buffer capacity describes the ability of a buffer to resist changes in pH when H^+ or OH^- is added. For quantitative purposes, it can be defined as the number of mol/L of H^+ or OH^- needed to change the pH of a buffer by 1 unit. In other words, BC is the change in the total $[H^+]$ or $[OH^-]$ per unit change in pH:

$$\text{BC} = \frac{d[\text{H}^+]_\text{T}}{d\text{pH}} = -\frac{d[\text{OH}^-]_\text{T}}{d\text{pH}} \tag{60}$$

where $d[H^+]_T$ is the change in the total H^+ concentration (i.e., bound plus free $[H^+]$) and is given by equation 66, below. To calculate BC, $d[H^+]_T$ is substituted by equation 66. When this substitution is made, equation 60 reveals that the capacity of any buffer is dependent on its concentration, maximal at the pK_a, and drops off to lower values as pH deviates from the pK_a. BC values outside the range $pH = pK_a \pm 1$ are too low for effective buffering, and buffers should not be used at pH values outside this range.

Often, a reaction in a buffered assay medium releases or takes up H^+. The total amount of H^+ released or taken up is represented by $d[H^+]_T$ and can be calculated using the Henderson-Hasselbach equation, or directly from equation 66 which is derived as follows:

When a small amount of H^+ is added to a buffer, most of the ions bind to the conjugate base (A^-) to form the conjugate acid (HA), and a small amount remains as free H^+. Therefore, [HA] and $[H^+]$ in the buffer have changed by the amounts Δ[HA] and $\Delta[H^+]$, respectively. This means that

$$d[\text{H}^+]_\text{T} = \Delta[\text{HA}] + \Delta[\text{H}^+]$$

Let the initial pH of the buffer be pH_1 and the final pH be pH_2. Then,

$$d[H^+]_T = ([HA_{pH_2}] - [HA_{pH_1}]) + ([H^+_{pH_2}] - [H^+_{pH_1}]) \tag{61}$$

From equation 58,

$$\frac{[A^-]}{[HA]} = 10^{(pH - pK_a)}$$

$$\therefore\ [A^-] = [HA] \times 10^{(pH - pK_a)} \tag{62}$$

Also,

$$[HA] + [A^-] = C \tag{63}$$

where C is the total molarity of the buffer. Substituting equation 62 into equation 63 yields:

$$[HA] + ([HA] \times 10^{(pH - pK_a)}) = C$$

$$[HA]\ (1 + 10^{(pH - pK_a)}) = C$$

$$\therefore\ [HA] = \frac{C}{1 + 10^{(pH - pK_a)}} \tag{64}$$

From equation 51,

$$pH = -\log [H^+]$$

$$\therefore\ [H^+] = 10^{-pH} \tag{65}$$

Substituting equations 64 and 65 into equation 61 yields:

$$d[H^+]_T = \left(\frac{C}{1 + 10^{(pH_2 - pK_a)}} - \frac{C}{1 + 10^{(pH_1 - pK_a)}} \right) + (10^{-pH_2} - 10^{-pH_1}) \tag{66}$$

Example: See problem 26 in section 2.4.

2.3. Preparation of Buffers and Related Topics

A. PREPARATION OF BUFFERS

Methods used to prepare buffers are based on the basic principle specified by the Henderson-Hasselbach equation: That the pH of a buffer is determined by the (conjugate base)/(conjugate acid) ratio, and vice versa. Preparing a buffer of a given pH, therefore, requires the preparation of a solution that contains the conjugate base and the conjugate acid at a ratio equal to the value predicted by the Henderson-Hasselbach equation. The appropriate ratio can be established by adding acid or base to a solution of one of the conjugates to adjust the pH to the desired value; or by dissolving precalculated quantities of the conjugates. The following three methods are based on these principles.

Method 1. Simple pH adjustment

(1) Calculate the weight of the buffer needed to prepare a solution of a given volume and molarity using equation 9 (use any of the conjugate forms of the buffer).

(2) Dissolve it in a volume of deionized H_2O, which is about eight-tenths of the desired final volume.

(3) Adjust the pH to the desired value, then add additional H_2O to adjust the volume to the final value.

Method 2. Using calculated amounts of the conjugate forms

(1) Use the Henderson-Hasselbach equation to calculate the concentrations of the conjugate acid and base that will be present in the buffer at the given pH.

(2) Use equation 9 to calculate the weight corresponding to each of the concentrations in step 1.

(3) Dissolve these quantities as in step 2 of method 1, above.

(4) Adjust the volume to the target value by adding additional H_2O.

(5) Check the pH. It should be within ±0.2 units of the expected value.

Examples: See problems 24a–d in section 2.4.

Method 3. Mixing proportionate volumes of the conjugate forms

(1) For each conjugate, calculate the weight needed to prepare a solution at a concentration equal to that of the buffer.

(2) Calculate the ratio $[A^-]/[HA]$ or $[R\text{-}NH_2]/[R\text{-}NH_3^+]$ at the given pH of the buffer. Use equations 55 or 56 (or equations 58 or 59).

(3) Use this ratio to calculate the volumes of both solutions which, when combined, yield a buffer with the given pH, concentration, and volume.

(4) Prepare the buffer by combining the measured volumes. Check the pH. It should be within ±0.2 units of the expected value.

Examples: See problem 25a–c in section 2.4.

Recipes for preparing common laboratory buffers based on Method 3 are given in Appendix H.

B. CHANGES IN BUFFER pH UPON DILUTION

When a concentrated buffer is diluted to prepare a working buffer, the pH of the dilute buffer changes even though no H^+ or OH^- has been added. The reasons for this change are manifold, but only the most important two are discussed below.

(1) Effect of dilution on activity coefficient

Some of the conjugate species in a buffer are ions. The effective molar concentration or activity of dissolved ions is defined as follows:

$$\text{Activity} = \gamma M \tag{67}$$

where γ and M are the activity coefficient and molarity of the ion, respectively. Substituting this into the Henderson-Hasselbach equation yields:

For a weak acid,

$$\text{pH} = \text{p}K_a + \log \frac{\gamma [\text{A}^-]}{[\text{HA}]} \tag{68}$$

For a weak base,

$$\text{pH} = \text{p}K_a + \log \frac{[\text{R-NH}_2]}{\gamma [\text{R-NH}_3^+]} \tag{69}$$

Because activity coefficients increase with increasing dilution, $\gamma [A^-]/[HA]$ or $[R\text{-}NH_2]/\gamma[R\text{-}NH_3^+]$ will change upon dilution, resulting in a change in the buffer's pH. For a weak acid, the pH increases because log ($\gamma[A^-]/[HA]$) increases, and for a weak base, the pH decreases because log ($[R\text{-}NH_2]/\gamma[R\text{-}NH_3^+]$) decreases with increasing dilution.

(2) Effect of dilution on the ionization of acids and bases

The degree of ionization of acids and bases increases as their concentration decreases with dilution. Thus, $[A^-]$ or $[R\text{-}NH_2]$ increases while $[HA]$ or $[R\text{-}NH_3^+]$ decreases. Log ($[A^-]/[HA]$) or log $[RNH_2]/[R\text{-}NH_3^+]$ increases accordingly, resulting in a pH increase upon dilution of a weak acid or a weak base buffer.

C. POLYPROTIC WEAK ACIDS

Polyprotic acids have more than one ionizable hydrogen. When dissolved, the hydrogen atoms dissociate sequentially as the pH of the solution increases, resulting in a stepwise profile in a pH titration curve (Figure 2.2). To illustrate this, the triprotic acid H_3PO_4 ionizes in three steps as follows:

$$H_3PO_4 \overset{K_{a_1}}{\rightleftarrows} H^+ + H_2PO_4^- \overset{K_{a_2}}{\rightleftarrows} H^+ + HPO_4^{2-} \overset{K_{a_3}}{\rightleftarrows} H^+ + PO_4^{3-}$$

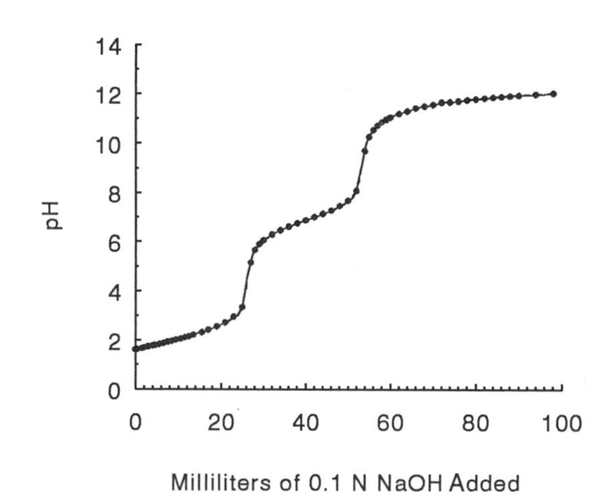

Figure 2.2. The pH titration curve for the triprotic weak acid, H_3PO_4.

The ionization equilibrium can be written in three steps as though three different weak acids were involved.

$$H_3PO_4 \rightleftarrows H^+ + H_2PO_4^-;\ K_{a_1} = \frac{[H^+][H_2PO_4^-]}{[H_3PO_4]} = 1.1 \times 10^{-2} \tag{71}$$

$$H_2PO_4^- \rightleftarrows H^+ + HPO_4^{2-};\ K_{a_2} = \frac{[H^+][HPO_4^{2-}]}{[H_2PO_4^-]} = 1.6 \times 10^{-7} \tag{72}$$

$$HPO_4^{2-} \rightleftarrows H^+ + PO_4^{3-};\quad K_{a_3} = \frac{[H^+][PO_4^{3-}]}{[HPO_4^{2-}]} = 4.8 \times 10^{-13} \tag{73}$$

The pK_a values are pK_{a_1}, 1.96; pK_{a_2}, 6.8; and pK_{a_3}, 12.3, corresponding to the acid ionization constants, K_{a_1}, K_{a_2}, and K_{a_3}, respectively. Accordingly, the pH titration curve (Figure 2.2) shows a three-step profile. Buffer calculations are done separately for each conjugate pair present in the buffer. Exact calculations are difficult because at any pH within the buffering range of the acid, all the conjugate forms are present and require separate calculations. To simplify the calculations, approximations are introduced by ignoring those conjugate forms with concentrations that are too low relative to the others. As a guide, when the concentration is less than 1/100th of the total buffer concentration, that conjugate form is ignored in the calculations.

Examples: See problems 24c, 24d, 25b, and 25c in section 2.4.

2.4. Practical Examples

PROBLEM 24

Prepare 0.800 L of 0.050 M solutions of the following buffers using calculated amounts of the conjugate acid and base. (a) Tris–acetate, pH 8.5 using Tris base (fw, 121.1) and 10.0 M acetic acid solution. (b) Acetate, pH 5.0 using sodium acetate trihydrate (fw, 136.08) and 10.0 M acetic acid solution. (c) Citrate (Cit), pH 5.0 using NaH_2Cit (fw, 214.1), Na_2HCit (fw, 236.1), and $Na_3Cit{\cdot}2H_2O$ (fw, 294). (d) Phosphate, pH 7.2 using $NaH_2PO_4{\cdot}H_2O$ (fw, 138) and $Na_2HPO_4{\cdot}H_2O$ (fw, 268.07). The pK_a values are: Tris, 8.1; acetic acid, 4.8; citric acid, 3.1, 4.8, and 5.4 for pK_{a_1}, pK_{a_2}, and pK_{a_3}, respectively; phosphoric acid, 1.96, 6.8, and 12.3 for pK_{a_1}, pK_{a_2}, and pK_{a_3}, respectively.

SOLUTION:

(a) Preparation of Tris–acetate buffer using method 2

At pH 8.5, Tris exists as the conjugate base, Tris-NH_2 and the conjugate acid, Tris-NH_3^+. Starting with Tris base which is 100% Tris-NH_2, Tris-NH_3^+ will be produced by adding acetic acid (HOAc) to convert Tris-NH_2 to Tris-NH_3^+ at a 1:1 equivalent ratio.

$$\therefore \ [\text{HOAc}] = [\text{Tris-NH}_3^+].$$

Calculate the following:

(i) The weight of Tris base needed; (ii) the [Tris-NH_3^+] at pH 8.5;
(iii) the milliliters of 10M HOAc needed. Using equation 9,

$$\text{g of Tris base needed} = \text{M} \times \text{mol wt} \times \text{L}$$

$$= 0.050 \text{ mol/L} \times 121.1 \text{ g/mol} \times 0.800 \text{ L}$$

$$= 4.8 \text{ g}$$

[Tris-NH_3^+] is calculated using the Henderson-Hasselbach equation (equation 56):

$$\text{pH} = \text{p}K_a + \log \frac{[\text{R-NH}_2]}{[\text{R-NH}_3^+]}$$

When solved as shown on page 36,

$$\frac{[\text{R-NH}_3^+]}{[\text{R-NH}_2]} = 10^{(\text{pH} - \text{p}K_a)}$$

Substituting 8.1 for pK_a and 8.5 for pH,

$$\frac{[\text{Tris-NH}_2]}{[\text{Tris-NH}_3^+]} = 10^{(8.5 - 8.1)} = 10^{0.4} = 2.51$$

$$\therefore\ [\text{Tris-NH}_2] = 2.51\ [\text{Tris-NH}_3^+] \tag{74}$$

Also,

$$[\text{Tris-NH}_2] + [\text{Tris-NH}_3^+] = 0.050\ \text{M}\ \ (\text{i.e, total [Tris]} = 0.050\text{M})$$

$$\therefore\ [\text{Tris-NH}_2] = 0.050 - [\text{Tris-NH}_3^+] \tag{75}$$

Combining equations 74 and 75,

$$2.51\ [\text{Tris-NH}_3^+] = 0.050 - [\text{Tris-NH}_3^+]$$

$$3.51\ [\text{Tris-NH}_3^+] = 0.050\ \text{M}$$

$$\therefore\ [\text{Tris-NH}_3^+] = \frac{0.050\ \text{M}}{3.51} = 0.014\ \text{M}$$

$$[\text{HOAc}]\ \text{in the buffer} = [\text{Tris-NH}_3^+] = 0.014\ \text{M}$$

V_1, the volume of the 10.0 M HOAc that will yield this concentration in the buffer, is calculated using equation 39:

$$V_1 = \frac{C_2 V_2}{C_1} = \frac{0.0143\ \text{M} \times 0.800\ \text{L}}{10.0\ \text{M}} = 0.00114\ \text{L} = 1.1\ \text{mL}$$

where V_2 is the final volume of the buffer; C_1 and C_2 are the concentrations of HOAc in the 10.0 M HOAc and the final buffer, respectively.

To prepare the buffer, dissolve 4.8 g of Tris base in 700 mL of H_2O, add 1.1 mL of 10.0 M HOAc, and adjust the volume to 800 mL with H_2O. The pH should be 8.5.

Note: HOAc (pK_a, 4.8) is completely ionized at pH 8.5 and does not contribute to the buffering capacity of the solution.

(b) Preparation of acetate buffer using method 2
The conjugate base is acetate (OAc^-) supplied by NaOAc, and the conjugate acid is acetic acid (HOAc). [OAc^-] and [HOAc] at pH 5.2 are calculated and used to calculate the grams of NaOAc and the milliliters of 10.0 M HOAc needed. The pK_a of HOAc is 4.8. Using the Henderson-Hasselbach equation (equation 55),

$$pH = pK_a + \log \frac{[OAc^-]}{[HOAc]}$$

When solved as shown on page 36,

$$\frac{[OAc^-]}{[HOAc]} = 10^{(pH - pK_a)}$$

Substituting 4.8 for pK_a and 5.0 for pH,

$$\frac{[OAc^-]}{[HOAc]} = 10^{(5.0 - 4.8)} = 10^{0.2} = 1.58$$

$$\therefore\ [OAc^-] = 1.58\,[HOAc] \qquad (76)$$

Also,

$$[OAc^-] + [HOAc] = 0.050\text{ M (i.e., total }[OAc^-] = 0.050\text{M)}$$

$$\therefore\ [OAc^-] = 0.050 - [HOAc] \qquad (77)$$

Combining equations 76 and 77,

$$1.58\,[HOAc] = 0.050 - [HOAc]$$

$$2.58\,[HOAc] = 0.050\text{ M}$$

$$\therefore\ [HOAc] = 0.050 \div 2.58 = 0.0194\text{ M}$$

and

$$[OAc^-] = (0.050 - 0.0194)\text{ M} = 0.0306\text{ M}$$

The grams of $NaOAc \cdot 3H_2O$ needed are calculated using equation 9,

$$\text{g of } NaOAc \cdot 3H_2O \text{ needed} = \text{M} \times \text{mol wt} \times \text{L}$$

$$= 0.0306\text{ mol/L} \times 136.08\text{ g/mol} \times 0.800\text{ L}$$

$$= 3.3\text{ g}$$

The volume of 10.0 M HOAc needed (V_1) is calculated using equation 39:

$$V_1 = \frac{C_2 V_2}{C_1} = \frac{0.0194 \text{ M} \times 0.800 \text{ mL}}{10.0} = 0.00155 \text{ L or } 1.6 \text{ mL}$$

where V_2 is the final volume of the buffer; C_1 and C_2 are the concentrations of HOAc in the 10.0 M HOAc and the final buffer, respectively.

To prepare the buffer, dissolve 3.3 g of NaOAc·3H_2O in 700 mL of H_2O, add 1.6 mL of 10.0 M HOAc, and adjust the volume to 800 mL with H_2O. The pH should be 5.0.

(c) Preparation of citrate buffer using method 2
Citric acid (H_3Cit) is triprotic; pK_{a_1}, 3.1; pK_{a_2}, 4.8; pK_{a_3}, 5.4. At pH 5.0, most of the H_3Cit has dissociated to H_2Cit^-, $HCit^{2-}$, and Cit^{3-}. [H_3Cit] is too low—about 1/100 of the total [citrate]—and is ignored in the calculations. The dissociation equilibrium is, therefore,

$$H_2Cit^- \overset{K_2}{\rightleftharpoons} HCit^{2-} \overset{K_3}{\rightleftharpoons} Cit^{3-}$$

Calculate the molarity of each conjugate form present in the buffer, and the grams of NaH_2Cit, Na_2HCit, and $Na_3Cit \cdot 2H_2O$ needed. Using the Henderson-Hasselbach equation (equation 55),

$$\text{pH} = \text{p}K_a + \log \frac{[HCit^{2-}]}{[H_2Cit^-]}$$

When solved as shown on page 36,

$$\frac{[HCit^{2-}]}{[H_2Cit^-]} = 10^{(\text{pH} - \text{p}K_{a_2})}$$

Substituting 4.8 for pK_{a_2}, and 5.0 for pH,

$$\frac{[HCit^{2-}]}{[H_2Cit^-]} = 10^{(5.0 - 4.8)} = 10^{0.2} = 1.58$$

$$\therefore [H_2Cit^-] = \frac{[HCit^{2-}]}{1.58} = 0.633\,[HCit^{2-}] \tag{78}$$

A similar calculation for the second equilibrium yields

$$[Cit^{3-}] = 0.398\,[HCit^{2-}] \tag{79}$$

Also,

$$[H_2Cit^-] + [HCit^{2-}] + [Cit^{3-}] = 0.050 \text{ M (i.e., total [Cit]} = 0.050 \text{ M)} \quad (80)$$

Substituting equations 78 and 79 into equation 80 yields:

$$0.633\ [HCit^{2-}] + [HCit^{2-}] + 0.398\ [HCit^{2-}] = 0.050 \text{ M}$$

$$2.031\ [HCit^{2-}] = 0.050 \text{ M}$$

$$\therefore\ [HCit^{2-}] = 0.0246 \text{ M}$$

Using equations 78 and 79, respectively,

$$[H_2Cit^-] = 0.0157 \text{ M},$$

and

$$[Cit^{3-}] = 0.00979 \text{ M}$$

The grams of citrate salts needed are then calculated using equation 9:

$$\text{g of } NaH_2Cit = \text{M} \times \text{mol wt} \times \text{L}$$

$$= 0.0157 \text{ mol/L} \times 214 \text{ g/mol} \times 0.800 \text{ L} = 2.7 \text{ g}$$

Similarly,

$$\text{g of } Na_2HCit \text{ needed} = 4.6 \text{ g}$$

and

$$\text{g of } Na_3Cit{\cdot}3H_2O \text{ needed} = 2.3 \text{ g}$$

To prepare the buffer, dissolve 2.7 g of NaH_2Cit, 4.6 g Na_2HCit, and 2.3 g of $Na_3Cit{\cdot}3H_2O$ in 700 mL of H_2O. Adjust volume to 800 mL. The pH should be 5.0.

(d) Preparation of phosphate buffer using method 2
Phosphoric acid is triprotic with pK_{a_1}, 1.96; pK_{a_2}, 6.8; and pK_{a_3}, 12.3. At pH 7.4, only $H_2PO_4^-$ and HPO_4^{2-} are present in significant concentrations; when calculated, $[H_3PO_4]$ and $[PO_4^{3-}]$ are only about 10^{-5} times the total phosphate concentration. They are, therefore, ignored in the calculations. The predominant equilibrium is:

$$H_2PO_4^- \xrightleftharpoons{K_2} HPO_4^{2-}$$

The conjugate acid, $H_2PO_4^-$, and the conjugate base, HPO_4^{2-}, are supplied by

NaH_2PO_4 and Na_2HPO_4, respectively. Calculate the $[H_2PO_4^-]$ and $[HPO_4^{2-}]$ at pH 7.4 and use the results to calculate the grams of NaH_2PO_4 and Na_2HPO_4 needed. Using the Henderson-Hasselbach equation (equation 55),

$$\text{pH} = \text{p}K_a + \log \frac{[HPO_4^{2-}]}{[H_2PO_4^-]}$$

When solved as shown on page 36,

$$\frac{[HPO_4^{2-}]}{[H_2PO_4^-]} = 10^{(\text{pH} - \text{p}K_{a_2})}$$

Substituting 6.8 for $\text{p}K_{a_2}$ and 7.4 for pH,

$$\frac{[HPO_4^{2-}]}{[H_2PO_4^-]} = 10^{(7.4-6.8)} = 10^{0.6} = 3.98$$

$$\therefore\ [HPO_4^{2-}] = 3.98\,[H_2PO_4^-] \qquad (81)$$

Also,

$$[HPO_4^{2-}] + [H_2PO_4^-] = 0.050\text{ M (i.e., total [phosphate]} = 0.050\text{ M)}$$

$$[HPO_4^{2-}] = 0.050 - [H_2PO_4^-] \qquad (82)$$

Combining equations 81 and 82 yields:

$$3.98\,[H_2PO_4^-] = 0.050 - [H_2PO_4^-]$$

$$4.98\,[H_2PO_4^-] = 0.050$$

$$[H_2PO_4^-] = \frac{(0.050)}{4.98} = 0.0100\text{ M}$$

and

$$[HPO_4^{2-}] = (0.050 - 0.0100) = 0.0400\text{ M}$$

The grams of phosphate salts needed are calculated using equation 9:

$$\text{g of } NaH_2PO_4{\cdot}H_2O \text{ needed} = \text{M} \times \text{mol wt} \times \text{L}$$

$$= 0.0100\text{ mol/L} \times 138\text{ g/mol} \times 0.800\text{ L}$$

$$= 1.1\text{ g}$$

Similarly,

g of $Na_2HPO_4 \cdot 7H_2O$ needed $= 0.0400$ mol/L × 268 g/mol × 0.800 L

$= 8.6$ g

Therefore, dissolve 1.1 g of NaH_2PO_4 and 8.6 g of Na_2HPO_4 in 700 mL of H_2O and adjust volume to 800 mL. The pH should be 7.4.

PROBLEM 25

Prepare 1.00 L of 0.100 M solutions of the following buffers by mixing appropriate volumes of the conjugate acid and base: (a) Acetate, pH 5.0, using sodium acetate trihydrate (fw 136.08) and concentrated acetic acid. (b) Phthalate, pH 6.0, using KH(phthalate) (fw 204.22) and K_2(phthalate) (fw 242.32). (c) Phosphate, pH 7.4, using $NaH_2PO_4 \cdot H_2O$ (fw 137.99) and $Na_2HPO_4 \cdot 7H_2O$ (fw 268.07). The pK_a values are acetic acid, 4.8; phthalic acid, 2.95 and 5.41; phosphoric acid, 1.96, 6.8, and 12.3.

SOLUTION:

For each buffer, prepare 1.00 L of a 0.100 M solution of the conjugate base and acid separately. Calculate the [conjugate base]/[conjugate acid] ratio. Use this ratio to calculate the volumes of the conjugate acid and conjugate base needed to prepare 1 L of each buffer.

(a) Preparation of acetate buffer using method 3
The conjugate base is acetate (OAc^-), and the conjugate acid is acetic acid (HOAc). Using equation 9,

g of $NaOAc \cdot 3H_2O$ required for 1.00 L of a 0.100 M solution

$= \text{M} \times \text{mol wt} \times \text{L}$

$= 0.100 \text{ mol/L} \times 136.08 \text{ g/mol} \times 1.00 \text{ L} = 13.6 \text{ g}$

V_1, the volume of concentrated HOAc needed to prepare 1.00 L of the 0.100 N solution, is calculated using equation 39:

$$V_1 = \frac{C_2V_2}{C_1} = \frac{0.100 \text{ N} \times 1.00 \text{ L}}{17.4 \text{ N}}$$

$= 5.747 \times 10^{-3}$ L or 5.75 mL.

where V_2 is the final volume of the 0.100 N HOAc; C_1 and C_2 are the concentrations of HOAc in the concentrated and the 0.100 N acid solutions, respectively. **Prepare the solutions using these calculated amounts.** Calculate the [OAc^-]/[HOAc] ratio using equation 55:

$$pH = pK_a + \log \frac{[OAc^-]}{[HOAc]}$$

When solved as shown on page 36, with $pK_a = 4.8$, and pH = 5.0,

$$\frac{[OAc^-]}{[HOAc]} = 10^{(pH - pK_a)} = 10^{(5.0 - 4.8)} = 10^{0.2} = 1.58$$

This means that 1.58 parts of OAc^- is present for every part of HOAc, a total of 2.58 parts. Thus,

1 part of 1.00 L buffer = (1000 mL)/2.58 = 387.6 mL

$\therefore$ vol of 0.100 M HOAc needed = 388 mL

and

vol of 0.100 M $NaOAc^-$ needed = 387.6 mL × 1.58

= 612 mL

$\therefore$ **Mix 388 mL of the 0.100 M HOAc and 612 mL of the 0.100 M NaOAc to yield 1.00 L of the 0.100 M acetate buffer, pH 5.0.**

(b) Preparation of phthalic buffer using method 3
First, calculate the grams of KH(phthalate) and K_2(phthalate) needed to prepare 1.00 L of a 0.100 M solution of each (equation 9):

g of KH(phthalate) needed = M × mol wt × L

= 0.100 mol/L × 204.22 g/mol × 1.00 L

= 20.4 g

Similarly,

g of K_2(phthalate) needed = 0.100 mol/L × 242.32 g/mol × 1.00 L

= 24.2 g

Prepare the solutions using these calculated amounts. Next, calculate the $[pht^{2-}]/[Hpht^-]$ ratio: Phthalic acid (H_2pht) is diprotic; pK_{a_1}, 2.95; pK_{a_2}, 5.41. It dissociates in two steps:

$$H_2pht \xrightleftharpoons{K_{a_1}} Hpht^- \xrightleftharpoons{K_{a_2}} pht^{2-}$$

At pH 6.0, most of the H_2pht has dissociated to $Hpht^-$. The remaining $[H_2pht]$ is too low and is ignored in the calculations. The $[pht^{2-}]/[Hpht^-]$ ratio is calculated using the Henderson-Hasselbach equation (equation 55):

$$pH = pK_a + \log \frac{[pht^{2-}]}{[Hpht^-]}$$

When solved as shown on page 36, with $pK_{a_2} = 5.41$, pH = 6.0,

$$\frac{[pht^{2-}]}{[Hpht^-]} = 10^{(pH - pK_{a_2})} = 10^{(6.0 - 5.41)} = 10^{0.59} = 3.89$$

This means that 3.89 parts of pht^{2-} is present for every part of $Hpht^-$, a total of 4.89 parts.

1 part of 1.00 L buffer = (1000 mL)/4.89 = 204.5 mL

∴ vol of 0.100 M KHpht needed = 204 mL

and

vol of 0.100 M K_2pht needed = 204.5 mL × 3.89 = 796 mL

∴ **Mix 204 mL of the 0.100 M KHpht and 796 mL of the 0.100 M K_2pht to yield 1.00 L of the 0.100 M phthalate buffer, pH 6.0.**

(c) Preparation of phosphate buffer using method 3
First, use equation 9 to calculate the grams of NaH_2PO_4 and Na_2HPO_4 needed to prepare 1.00 L of a 0.100 M solution of each:

g of $NaH_2PO_4 \cdot H_2O$ needed = M × mol wt × L

= 0.100 mol/L × 138 g/mol × 1.00 L

= 13.8 g

Similarly,

g of $Na_2HPO_4 \cdot 7H_2O$ needed = 0.100 mol/L × 268 g/mol × 1.00 L

= 26.8 g

Prepare the 0.100 M solutions using these calculated amounts. Next, calculate the $[HPO_4^{2-}]/[H_2PO_4^-]$ ratio: H_3PO_4 dissociates in three steps, and at pH 7.4, the predominant conjugate acid is $H_2PO_4^-$, and the predominant conjugate base is HPO_4^{2-} (see problem 24d for details). Calculate the $[HPO_4^{2-}]/[H_2PO_4^-]$ ratio using the Henderson-Hasselbach equation (equation 55),

$$\text{pH} = \text{p}K_a + \log \frac{[\text{HPO}_4^{2-}]}{[\text{H}_2\text{PO}_4^-]}$$

When solved as shown on page 36, with $\text{p}K_{a_2} = 6.8$, and pH 7.4,

$$\frac{[\text{HPO}_4^{2-}]}{[\text{H}_2\text{PO}_4^-]} = 10^{(\text{pH} - \text{p}K_{a_2})} = 10^{(7.4-6.8)} = 10^{0.6} = 3.98$$

This means that 3.98 parts of HPO_4^{2-} is present for every part of $H_2PO_4^-$, a total of 4.98 parts. Thus,

$$1 \text{ part of } 1.00 \text{ L buffer} = (1000 \text{ mL})/4.98 = 200.8 \text{ mL}$$

$$\therefore \text{ vol of } 0.100 \text{ M NaH}_2\text{PO}_4 \text{ needed} = 201 \text{ mL}$$

and

$$\text{vol of } 0.100 \text{ M Na}_2\text{HPO}_4 \text{ needed} = 200.8 \text{ mL} \times 3.98$$

$$= 799 \text{ mL}$$

To prepare the buffer, mix 201 mL of the 0.100 M NaH_2PO_4 and 799 mL of the 0.100 M Na_2HPO_4 to yield 1.00 L of the 0.100 M phosphate buffer, pH 7.4.

PROBLEM 26

The pH of a 2.0 mL assay medium buffered with 0.050 M glycylglycine (pK_a 8.4) decreased from 8.5 to 8.3 at the end of the assay. Calculate the moles of H^+ released into the medium.

SOLUTION:

Calculate the change in the total $[H^+]$ ($d[H^+]_T$) in the buffer due to the change in pH from 8.5 to 8.3. From equation 66,

$d[H^+]_T$ =

$$\frac{C}{1 + 10^{(pH_2 - pK_a)}} - \frac{C}{1 + 10^{(pH_1 - pK_a)}} + (10^{-pH_2} - 10^{-pH_1})$$

$$= \frac{0.050\ M}{1 + 10^{(8.3 - 8.4)}} - \frac{0.050\ M}{1 + 10^{(8.5 - 8.4)}} + (10^{-8.3} - 10^{-8.5})$$

where C, the total buffer concentration = 0.050 M, $pK_a = 8.4$, $pH_1 = 8.5$, and $pH_2 = 8.3$.

$$\therefore\ d[H^+]_T = (0.0279 - 0.0221)\ M + (1.85 \times 10^{-9}\ M) = 0.0058\ M$$

$$= (0.0279 - 0.0221)\ M = 0.0058\ M$$

H^+ released into the 2.0 mL assay medium =

$$\frac{0.0057\ mol}{1000\ mL} \times 2.0\ mL = 11.6 \times 10^{-6}\ mol\ \text{or}\ 12\ \mu mol$$

PROBLEM 27

What is the pH of (a) 0.050 M HCl, and (b) 0.100 M HCl.

SOLUTION:

In dilute solutions, strong acids and bases ionize completely. Therefore, $[H^+] = [HCl]$.

(a) $[H^+] = 0.0050$ M. Using equation 51,

$$pH = -\log [H^+] = -\log 0.0050 = 2.3$$

(b) $[H^+] = 0.100$

$$pH = -\log 0.100 = 1.0$$

Note: The pH of the solution depends on the concentration of the acid, and unlike weak acids (or weak bases), strong acids (or strong bases) do not possess buffering properties.

PROBLEM 28

A 0.200 L aliquot of 0.050 M solution of dibasic sodium phosphate (Na_2HPO_4) is mixed with 0.160 L of 0.050 M monobasic sodium phosphate (NaH_2PO_4). Calculate the pH of the resulting buffer.

SOLUTION:

The conjugate acid is NaH_2PO_4, and the conjugate base is Na_2HPO_4. Since the concentrations of the two solutions are equal,

$$\frac{[Na_2HPO_4]}{[NaH_2PO_4]} = \frac{0.200 \text{ L}}{0.160 \text{ L}} = 1.250$$

Using equation 55,

$$\begin{aligned} pH &= pK_a + \log \frac{[Na_2HPO_4]}{[NaH_2PO_4]} \\ &= 6.8 + \log 1.250 = 6.9 \end{aligned}$$

Chapter 3

SPECTROPHOTOMETRY: BASIC PRINCIPLES AND QUANTITATIVE APPLICATIONS

3.1. Principles and Techniques

Spectrophotometry is the science of measuring the light absorption characteristics of substances for use in determining their concentration, identity, and other properties. This technique is possible because many substances absorb light of specific wavelengths within the ultraviolet (200–400 nm), visible (400–700 nm), and near-infrared (700–1000 nm) regions of the electromagnetic spectrum, and transmit the remainder. The wavelength of the transmitted (or reflected) light is what imparts a characteristic color to a substance: For example, a red wine appears red because it transmits red light (wavelength, 600–700 nm), and absorbs light of shorter wavelengths.

A measure of the amount of light absorbed by a substance is called **absorbance** or **optical density (OD)**, and a measure of the amount transmitted is called **transmittance** (see Appendix A.3). The wavelength at which a substance has maximum absorbance is characteristic of the substance and is used to identify the substance. If two substances contain the same light-absorbing prosthetic group (e.g., riboflavin and flavin adenine dinucleotide both contain flavin), they usually absorb maximally at about the same wavelength, but the absorbance per mole of each substance **(molar extinction coefficient)** is usually different. A recording of the absorbance of a substance as a function of wavelength is called the **absorption spectrum** (Figure 3.1). The peaks (**absorption maxima**) on an absorbance spectrum are the regions of high absorption, and the troughs (**absorption minima**) are the regions of low absorption.

Absorbance is measured with a spectrophotometer, and the basics of this technique and the instrumentation are described in Appendix A.4. The rest of this chapter deals with basic calculations and applications of this versatile technique.

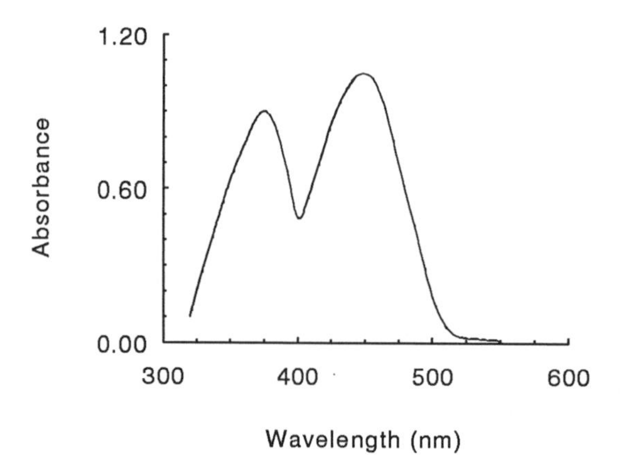

Figure 3.1. A spectrum of riboflavin in the visible and near-UV regions.

3.2. Quantitative Aspects and Basic Applications

A. THE BEER-LAMBERT LAW

This law is the basis for applying spectrophotometry to the calculation of concentrations. It states that the absorbance (A) of a substance in solution is linearly proportional to its concentration (C) and the length (ℓ) of the light path through the solution. The law is expressed mathematically (see Appendix A.3 for derivation) as follows:

$$A = E\ell C \tag{83}$$

Rearranging equation 83,

$$C = \frac{A}{E\ell} \tag{84}$$

and

$$E = \frac{A}{\ell C} \tag{85}$$

The proportionality constant, E, is characteristic of the substance at the wavelength in which the absorbance was measured. It is known by various other names: **extinction coefficient**, **absorption coefficient**, **absorptivity**, and **absorbency index**. Its unit and symbolic designation depend on the units of C and ℓ. When C and ℓ are 1 M and 1 cm, respectively, E is designated $E_{1cm}^{1M}(\lambda)$ or $\in$, the molar extinction coefficient; otherwise, the general designation used is $E_{\ell}^{C}(\lambda)$. Equation 84 is used to directly calculate the concentration of a substance once A, E, and ℓ are known. Equation 85 is used to calculate E for a substance at a given wavelength. To do this, the A of a known concentration of the substance is measured at this wavelength, using a cuvette with a known ℓ. The A, C, and ℓ are then used to calculate the E. The A can, alternatively, be obtained from a spectrum of a known concentration of the substance.

B. CALCULATING CONCENTRATIONS

(1) Nonenzymatic analytes

The Beer-Lambert law is used to calculate the concentration of substances using absorbance data and equation 84. The calculation is handled in two ways:

(a) Direct calculation

This is applicable to substances that directly absorb light. To calculate the concentration of a sample, first measure its absorbance. Obtain E and ℓ; if unknown, measure E as described above, and measure ℓ as the length of the cuvette's light path (usually 1 cm). Use equation 84 to calculate the concentration.

Examples: See problems 29, 30a, 32a, and 33 in section 3.3.

(b) Use of a standard curve to calculate concentration

The use of a standard curve is often necessary when analyzing substances that do not absorb light directly. These substances may be converted to light-absorbing species by either a chemical reaction or by coupling to other reactions in which a light absorber is being consumed or produced. Direct calculation of concentration in these cases is tedious and sometimes impossible because the extinction coefficients of the light-absorbing species are unknown. Standards of the pure substance are assayed along with the samples, and a standard curve is constructed by plotting the absorbance values for these standards against the corresponding concentrations. The concentration of the sample is then obtained by reading from this curve the concentration corresponding to the sample's absorbance. If the sample was diluted, the result is multiplied by the dilution factor. However, if the standards and the sample were diluted equally, the concentration of the sample should not be multiplied by the dilution factor since this factor was incorporated into the standard curve.

Example: See problem 34 in section 3.3.

Alternatively, the absorbance and concentration of a single standard can be used to calculate the unknown concentration, provided that this standard is within the linear range of the assay.

Example: See problem 34 in section 3.3.

A standard curve can also be used when several samples are being assayed simultaneously; reading the concentrations off the standard curve is faster than calculating them.

(2) Analyzing enzymes spectrophotometrically

The concentration of an enzyme is usually measured as activity. Because activity is defined as the amount of substrate consumed or product liberated per unit time, the activity of an enzyme can be determined spectrophotometrically if the substrate or product absorbs light, or if it can be coupled to a reaction in which a light absorber is being consumed or produced. Enzymes that catalyze oxidation-reduction (redox) reactions (e.g., the cytochromes), contain light-absorbing prosthetic groups. Their redox transition activities are best assayed spectrophotometrically (*4, 5*).

Examples: See problems 32, 33, 35, and 37, and Chapter 4 for the calculation of enzyme activities.

Notes: (a) The Beer-Lambert equation is linear and has zero intercept and a slope of $E\ell$; therefore, the standard curve should pass through the zero origin. (b) At higher concentrations, the curve usually becomes nonlinear mainly because of intermolecular interactions, depletion of light, light scattering, instrument limitations at low levels of transmitted light, and reagent depletion. Absorbance values above the linear limit do not give accurate concentrations. Therefore, **(i)** only the linear portion of a standard curve should be used to obtain the concentration of samples; samples with higher absorbance values should be diluted and reassayed. **(ii)** Except when a sophisticated spectrophotometer with high accuracy is used, absorbance values greater than about 1.5 should not be used for calculations; the sample should be diluted and reassayed.

C. QUANTIFYING NUCLEIC ACIDS

Nucleic acids absorb UV light between 250 and 280 nm, with DNA and RNA absorbing optimally at 260 nm (see Appendix E for specifics). As such, they can be quantified spectrophotometrically using equation 83. However, molecular biologists usually quantify nucleic acids in **absorbance units (AU)** or **optical density (OD)** units. One AU or OD unit of a substance (e.g., nucleic acid) is defined as the concentration that gives 1 AU at a given wavelength. The quantitative basis for this definition is obtained by rearranging equation 83:

$$\text{OD} = E\ell C$$

By rearranging and substituting 1 cm for ℓ:

$$\frac{\text{C}}{\text{OD}} = \frac{1}{E} = 1 \text{ AU} \tag{86}$$

C/OD defines 1 AU or 1 OD unit. Thus, **1 AU is the inverse of the absorption coefficient.** Equation 86 is used to obtain the value of 1 AU when the absorption coefficient is known, and vice versa. To calculate the concentration of nucleic acids (or other light-absorbing substances) using their AU and measured OD, the following equations can be used:

From equation 86,

$$\frac{C}{\text{OD}} = 1\ \text{AU}$$

$$\therefore\ C = 1\ \text{AU} \times \text{OD}$$

If the sample was diluted before measuring the OD and DF is the dilution factor, then

$$C = 1\ \text{AU} \times \text{OD} \times \text{DF} \qquad (87)$$

Note: The unit for concentration will be that of the AU, usually mg/mL.

Values for the AU and absorption coefficient for RNA and DNA are given in Table3.1, below. Additional data are given in Appendix E. To use these values to calculate concentrations, apply equation 87 when AU is known or equation 83 when the absorption coefficient is known.

Examples: See problems 30, 31, and 43 in section 3.3.

Base pairing and stacking decrease the absorbance of DNA and RNA. Thus, when double-stranded DNA melts, its absorbance increases [**hyperchromic effect**, (*6*)]. For this reason, double-stranded DNA (or RNA) has a lower absorption coefficient and a higher value for AU than the single-stranded forms.

The values for absorption coefficient and AU for short single-stranded oligonucleotides vary, depending on the length and base composition: The shorter the oligo, the higher the absorption coefficient.

TABLE 3.1. Absorption Constants for Selected Nucleic Acids[a]

Nucleic Acid	$E_{1cm}^{1mg/mL}$ (260 nm)	1 AU_{260} (μg/mL)
Double-stranded DNA	20	50
Single-stranded DNA or RNA[b]	25	40
Single-stranded oligos[c]	≈ 30	≈ 33
Single-stranded oligos[d]	≈ 40	≈ 25

[a] See Appendix E for more data.
[b] Greater than 100 nucleotides.
[c] 60-100 nucleotides.
[d] Less than 40 nucleotides.

When quantifying nucleic acids, take OD readings at 260 nm and 280 nm wavelengths. Calculate the concentration using the OD_{260}, then calculate the OD_{260}/OD_{280} ratio and use it to assess the purity of the sample: Pure DNA and RNA have OD_{260}/OD_{280} ratio of 1.8 and 2.0, respectively. If the ratios are significantly less than these values, then the samples are contaminated.

D. IDENTIFYING SUBSTANCES

All direct absorbers have characteristic absorption spectra (*7*). The spectrum may be regarded as an optical "fingerprint" which identifies the compound. Identification is made using important features of the spectrum: **(i)** The wavelengths at which the absorption peaks and troughs occur are characteristic of the substance, **(ii)** the extinction coefficients at the absorption maxima are also characteristic, and **(iii)** the ratio of the absorbance at two characteristic wavelengths is a constant. As an example, the spectrum of riboflavin (Figure 3.1) shows absorption maxima at 450 and 375 nm, and an absorption minima at 400 nm. The E_{1cm}^{1M} (450 and 374 nm) are always 1.22×10^4 and 1.06×10^4, respectively, and the A_{450}/A_{375} ratio is always 1.15 for pure riboflavin. Together, these characteristics distinguish riboflavin from other compounds, including flavin adenine dinucleotide (FAD) which also contains flavin, the light-absorbing prosthetic group in riboflavin. Other compounds can be identified by converting them to light-absorbing analogues, and then recording and analyzing their spectra.

E. CHARACTERIZING SUBSTANCES

Spectrophotometric techniques have been used to characterize biomolecules. Examples include the determination of the pK_a of weak acids (*8*), oxidation-reduction transitions of heme proteins (*4*), ligand-receptor interactions including enzyme substrate interactions (*7*), and transitions of DNA between single- and double-stranded forms (*6*).

3.3. Practical Examples

PROBLEM 29

The absorbance of a solution of riboflavin in a 1.0 cm cuvette was 0.610 at 450 nm and 0.530 at 375 nm. Calculate the concentration of riboflavin in the solution. The $E_{1\text{cm}}^{1\text{M}}$ of riboflavin at 450 nm and 375 nm are 1.22×10^4 and 1.06×10^4, respectively.

SOLUTION:

The concentration is calculated using either the A and E at 450 nm or the A and E at 375 nm. From equation 84,

$$C = \frac{A}{E\ell}$$

Using the data at 450 nm,

$$C = [\text{riboflavin}] = \frac{0.610}{1.22 \times 10^4\ \text{M}^{-1}\ \text{cm}^{-1} \times 1.0\ \text{cm}} = 5.0 \times 10^{-5}\ \text{M}$$

Using the data at 375 nm

$$C = [\text{riboflavin}] = \frac{0.530}{1.06 \times 10^4\ \text{M}^{-1}\ \text{cm}^{-1} \times 1.0\ \text{cm}} = 5.0 \times 10^{-5}\ \text{M}$$

PROBLEM 30

(a) Estimate the amount of DNA in a 20.0 mL sample if the absorbance at 260 nm of 10.0 μL of the sample in 1.99 mL of H_2O is 0.500. Use an $E_{1\text{cm}}^{1\ \text{mg/mL}}$ (260 nm) of 20.3 for DNA; light path, 1.0 cm. (b) An aliquot of a 2.0 mL solution of a purified protein was diluted 10-fold before 0.10 mL was mixed with 5.0 mL of Bradford reagent. If this gives an absorbance of 0.45 at 595 nm, estimate the milligrams of protein in the 2.0 mL solution given that 0.10 mL of a standard (1.0 mg/mL), similarly treated, gave a 0.510 absorbance.

SOLUTION:

(a) From equation 84,

$$C = \text{mg DNA/mL} = A/E\ell$$

$$\text{mg DNA/mL} = \frac{0.500}{20.3\ (\text{mg/mL})^{-1}\ \text{cm}^{-1} \times 1.0\ \text{cm}}$$

$$= 0.0246\ \text{mg/mL}$$

$$= \text{concn of DNA in the assay medium}$$

To calculate mg DNA/mL of sample, let V_1 be the volume of the sample (10.0 μL) added to the H_2O, and C_1 its DNA concentration; V_2 and C_2 are the final volume of the assay medium and the concentration of DNA in it, respectively. Using equation 39,

$$C_1 = \frac{C_2V_2}{V_1} = \frac{0.0246\ \text{mg/mL} \times 2.0\ \text{mL}}{0.0100\ \text{mL}} = 4.93\ \text{mg/mL}$$

This means that 1 mL of the sample contains 4.93 mg of DNA

$$\therefore\ \text{20 mL of sample contains}\ \frac{4.93\ \text{mg}}{1\ \text{mL}} \times 20.0\ \text{mL} = 99\ \text{mg}$$

(b) Since both the sample and the standard were treated similarly, the milligrams of protein in the assay medium can be calculated by proportion:

$$\frac{\text{absorbance of standard}}{\text{mg/mL of standard}} = \frac{\text{absorbance of sample}}{\text{mg/mL of sample}}$$

$$\text{mg/mL of sample} = \frac{\text{absorbance of sample} \times \text{mg/mL of standard}}{\text{absorbance of standard}} \times 10$$

$$\text{mg of sample} = \frac{0.45 \times 1.0\ \text{mg/mL}}{0.510} \times 10 = 8.82\ \text{mg/mL}$$

$$\therefore\ \text{mg of protein in the 2 mL sample} = 8.82\ \text{mg/mL} \times 2.0\ \text{mL}$$

$$= 18\ \text{mg}$$

PROBLEM 31

You received a vial containing a 0.20 mL solution of a 20-base single-stranded DNA. If the concentration of DNA is given as 0.450 OD_{260} units, convert this concentration to (a) µg/mL, (b) pM. (c) Calculate the total amount of DNA in pmol. For a 20-base single-stranded DNA, assume that 1 $AU_{260} \approx 25$ µg DNA/mL. (See also problem 42.)

SOLUTION:

(a) Using equation 89,

$$\text{concn of DNA} = 1\text{ AU} \times \text{OD} \times \text{DF}$$

$$= 25\ \mu\text{g/mL} \times 0.450 \times 1$$

$$= 11.2\ \mu\text{g DNA/mL}$$

(b) Molecular mass of the DNA ≈ 325 D/base × 20 bases = 6500 D. Using equation 10,

$$M = \frac{\text{wt}}{\text{mol wt} \times \text{L}}$$

$$= \frac{1.125 \times 10^{-5}\text{ g}}{6500\text{ g/mol} \times 2.0 \times 10^{-4}\text{ L}}$$

$$= 8.65 \times 10^{-6}\text{ mol/L}$$

$$= 8.65 \times 10^{-6}\text{ mol/L} \times \frac{1\ \mu\text{M}}{1 \times 10^{-6}\text{ mol/L}}$$

$$= 8.7\ \mu\text{M}$$

(c) Using equation 9,

$$M = \text{mol/L}$$

$$\therefore \text{ Total moles of DNA} = M \times L$$

$$= 8.65 \times 10^{-6}\text{ mol/L} \times 2.0 \times 10^{-4}\text{ L}$$

$$= 1.73 \times 10^{-9}\text{ mol}$$

$$= 1.73 \times 10^{-9}\text{ mol} \times \frac{1\text{pmol}}{1 \times 10^{-12}\text{ mol}}$$

PROBLEM 32

Cytochrome *o* oxidizes ubiquinol-1 (Q_1H_2) to ubiquinone-1 (Q_1) which absorbs light at 262 nm. After adding 80.0 µg of cytochrome *o* to an assay medium containing a saturating concentration of Q_1H_2, the absorbance changed from 0.010 to 0.190 in 2.0 min. Calculate (a) moles of Q_1H_2 oxidized, (b) the specific activity of cytochrome *o*. Total volume of the assay medium was 2.0 mL, E_{1cm}^{1mM} (262 nm) of Q_1 is 15.0, and ℓ is 1.0 cm. Oxidized is abbreviated ox.

SOLUTION:

(a) The moles of Q_1H_2 oxidized equal the moles of Q_1 formed, and is represented by the change in absorbance (ΔA). Using equation 84,

$$C = [Q_1H_2]_{ox} = \frac{\Delta A}{E\ell} = \frac{(0.190 - 0.010)}{15.0\ \text{mM}^{-1}\ \text{cm}^{-1} \times 1.0\ \text{cm}}$$

The total amount of oxidized Q_1H_2 in the 2.0 mL assay medium

$$= \frac{(0.190 - 0.010)}{15.0\ (\text{mmol/L})^{-1}\ \text{cm}^{-1} \times 1.0\ \text{cm}} \times \frac{2.0\ \text{mL}}{1000\ \text{mL/L}}$$

$$= 2.4 \times 10^{-5}\ \text{mmol or 24 nmol}$$

(b) Using equation 98,

$$\text{sp act. of cytochrome } o = \frac{\text{amount of } Q_1H_2 \text{ oxidized}}{\text{min} \times \text{mg of protein}}$$

$$= \frac{24\ \text{nmol}}{2.0\ \text{min} \times 0.0800\ \text{mg protein}}$$

$$= 150\ \text{nmol min}^{-1}\ \text{mg}^{-1}\ \text{protein}$$

PROBLEM 33

A 0.10 mL aliquot of bacterial membrane preparation decreased the 340 nm absorbance of NADH from 0.415 to 0.138 in 5.0 min, in a 2.0-mL assay medium. Calculate the specific and total activities of NADH dehydrogenase in 100.0 mL of the extract containing 2.0 mg of protein/mL. Assume a saturating concentration of NADH, and a 1.0 cm light path. The E_{1cm}^{1M} (340 nm) of NADH is 6220.

SOLUTION:

The change in absorbance (ΔA) represents the total amount of NADH oxidized. By using equation 84,

$$C = [\text{NADH}]_{\text{ox}} = \frac{\Delta A}{E\ell} = \frac{(0.415 - 0.138)}{6220\ \text{M}^{-1}\ \text{cm}^{-1} \times 1.0\ \text{cm}}$$

The total amount of oxidized NADH in the 2.0 mL assay medium

$$= \frac{(0.415 - 0.138)}{6220\ \text{M}^{-1}\ \text{cm}^{-1} \times 1.0\ \text{cm}} \times \frac{2.0\ \text{mL}}{1000\ \text{mL/L}}$$

$$= 8.91 \times 10^{-8}\ \text{mol or } 0.0890\ \mu\text{mol}$$

Total protein in the assay medium

$$= 2.0\ \text{mg/mL} \times 0.10\ \text{mL}$$

$$= 0.20\ \text{mg}$$

Using equation 93,

$$\text{sp act.} = \frac{0.0891\ \mu\text{mol of NADH}}{5.0\ \text{min} \times 0.20\ \text{mg of protein}}$$

$$= 0.089\ (\mu\text{mol min}^{-1})\ \text{mg}^{-1}\ \text{of protein}$$

Total protein/100.0 mL extract = 2.0 mg/mL × 100.0 mL = 200 mg

∴ Total activity in 100.0 mL extract

$$= 0.0891\ (\mu\text{mol min}^{-1})\ \text{mg}^{-1}\ \text{protein} \times 200\ \text{mg of protein}$$

$$= 18\ \mu\text{mol/min or } 18\ \text{IU}.$$

where 1 International Unit (IU) = 1 μmol/min (see page 73).

PROBLEM 34

p-Nitrophenol (PNP) absorbs light at 405 nm. A 0.10 mL aliquot of a sample containing PNP was added to 1.90 mL of H_2O and the absorbance measured at 405 nm was 0.550. The absorbance for standards treated similarly are as shown below. (a) What is the concentration of PNP in the sample? (b) If only the 30.0 µM standard is assayed, calculate the concentration of PNP in the sample. Assume that the extinction coefficient of PNP is unknown.

µM of standard	0.0	10.0	20.0	30.0	40.0
OD	0.000	0.180	0.421	0.592	0.810

SOLUTION:

(a) Plot the µM PNP versus OD (Figure 3.2). The concentration of PNP in the sample is the µM PNP on the standard curve corresponding to 0.550 AU. This value is 27.5 µM.

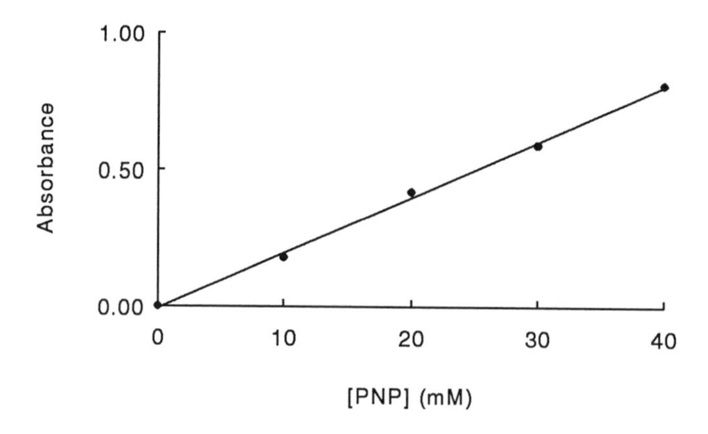

Figure 3.2. OD versus concentration standard curve for PNP.

(b) Since the 30.0 µM standard is within the linear range of the assay (see standard curve), it can be used to calculate the [PNP] in the sample by simple proportion:

$$\frac{A_{sample}}{C_{sample}} = \frac{A_{standard}}{C_{standard}}$$

$$C_{sample} = \frac{A_{sample} \times C_{standard}}{A_{standard}} = \frac{0.550 \times 30.0\ \mu M}{0.592} = 27.9\ \mu M$$

Chapter 4

ENZYME ASSAYS AND ACTIVITY

4.1. Introduction

Every enzyme assay involves the use of an enzyme as a catalyst to convert a substrate (and co-substrate in some cases) to product. To detect the reaction, a property (e.g., light absorption) of either the substrate, product, or co-substrate is measured at appropriate times. If none of them has a suitable property, the reaction can be coupled to a second reaction system that has a readily measurable property. Coupling is achieved by making a product of the first reaction serve as a substrate or co-substrate for the second. The stoichiometries of the reactions facilitate calculating the amount of substrate or product catalyzed in the first reaction.

Enzyme assays may be qualitative or quantitative. The qualitative assay seeks to establish the presence or absence of either the enzyme or the substrate in a sample. In contrast, the quantitative assay measures the amount of enzyme or substrate in a sample and requires optimal conditions if accuracy is desired. Quantitative assays can be classified into two broad categories: One category includes those that are aimed at determining the enzyme's activity or kinetic constants, and the other category includes those that are aimed at determining the concentration of substrate, using the enzyme as a reagent.

Applications for enzyme assays include the determination of the activity and kinetic constants for enzymes; determination of substrate or co-substrate concentration; determination of antigen concentration by means of enzyme immunoassays; qualitative detection of biomolecules, cellular components, cells, tissues, or other matrices in which a suitable enzyme or substrate serves as a marker.

The early parts of this chapter outline the quantitative basis for establishing conditions for enzyme assays. The later parts discuss how the raw data obtained from these assays are used to calculate activities and other parameters for enzymes. Definitions of activities, activity units, and kinetic constants are summarized.

4.2. Conditions Required for Quantitative Enzyme Assays

A. THE MICHAELIS-MENTON EQUATION AS A BASIS FOR ASSAY CONDITIONS

The Michaelis-Menton equation (*5, 7* for the derivation) expresses the quantitative aspects of enzyme kinetics. For an enzyme reaction,

$$v_i = \frac{[S]v_{max}}{K_m + [S]} \tag{88}$$

where v_i and v_{max} are the initial and maximal velocities, respectively, of the reaction; $[S]$ is the substrate concentration at the time v_i is measured; and K_m is the Michaelis constant, defined as the substrate concentration at half maximal velocity. The equation relates the initial velocity of the reaction to the maximal velocity and substrate concentration. If the true initial velocities are measured, then $[S] = [S_0]$ where $[S_0]$ is the initial substrate concentration. A plot of v_i versus $[S_0]$ illustrates this relationship (Figure 4.1). In the figure, the kinetics in region **A** form the basis for conditions required by assays that determine an enzyme's activity. Likewise, the kinetics in region **B** form the basis for conditions required by assays that determine the concentration of a substrate or co-substrate, using an enzyme as a reagent.

B. ASSAYS THAT MEASURE ENZYME ACTIVITY

In region **A** of Figure 4.1, v_i does not change as $[S_0]$ changes. When the rate of a reaction is independent of the concentrations of the reactants, the reaction has **zero-order** kinetics. Thus, reactions in region **A** have zero-order kinetics. Also, $[S_0] \gg K_m$, and the $[S]$ in the denominator of equation 88 is, therefore, far larger than K_m. The latter can then be ignored in the equation. In quantitative terms,

$$K_m + [S] \approx [S]$$

and equation 88 becomes:

$$v_i = \frac{[S]}{[S]} v_{max} = v_{max} \tag{89}$$

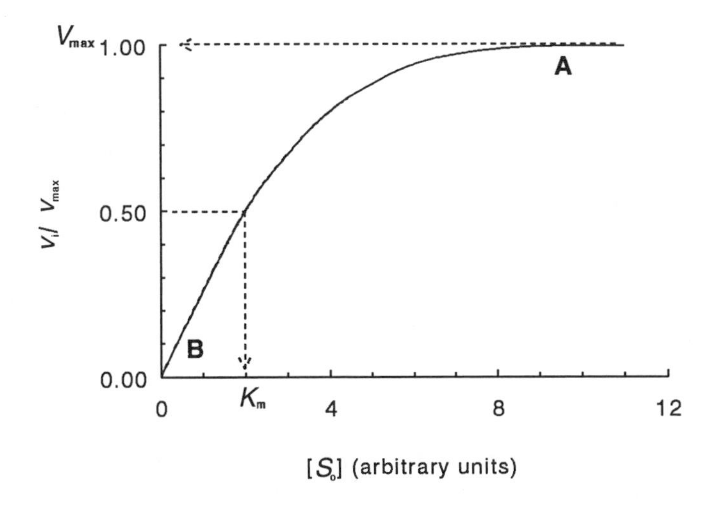

Figure 4.1. Relationship between initial velocity and initial substrate concentration of an enzyme-catalyzed reaction.

This means that in region **A**, the rate at which an enzyme converts its substrate to product (P) is maximal and constant. Consequently, as the reaction proceeds, $[S]$ decreases while $[P]$ increases linearly. When either is plotted against time (t), the slope of the curve yields the activity of the enzyme (Figure 4.2). In general, these properties of region **A** form the criteria for all quantitative assays that measure the optimal activities of enzymes. These criteria are summarized below:

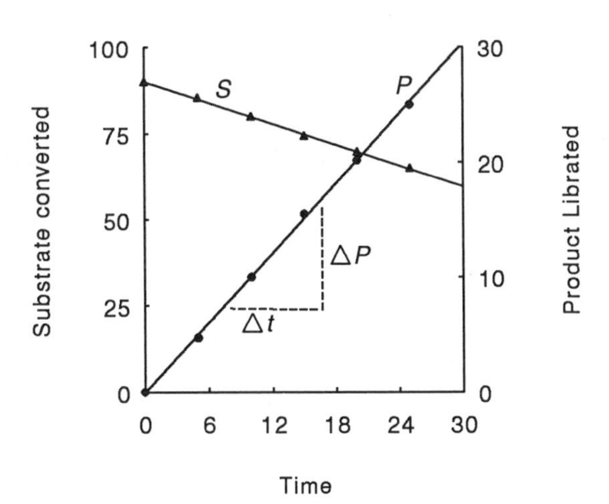

Figure 4.2. A plot of $[S]$ or $[P]$ versus time (t) when $[S]$ is saturating. DP/Dt is a measure of the reaction's velocity (v).

(1) $[S]$ should be saturating ($[S] \approx 100 \times K_m$ is suggested).
(2) The enzyme concentration used should give a linear product formation or substrate consumption over a time period sufficiently long to permit making accurate measurements.
(3) To verify that the above conditions are met, $[S]$ or $[P]$ is plotted against t, and the curve should be linear. Alternatively, v_i is plotted against t, and this curve should be a straight line parallel to the time axis. Also, a doubling of the enzyme concentration should result in a doubling of v_i. **Example:** See problem 37 in section 4.5.
(4) In addition to the above criteria specified by the Michaelis-Menton kinetics, the temperature and pH of the assay must be optimized because a given enzyme catalyzes optimally at a certain pH and temperature peak or within certain pH and temperature ranges.

The required enzyme concentration should be determined by assaying several dilutions of the enzyme using a fixed saturating $[S]$ (see problem 37). If K_m is unknown, the saturating $[S]$ should also be estimated by assaying several $[S]$ using a fixed adequate enzyme concentration. When v_i is plotted against $[S_0]$ in such a case, the saturating $[S_0]$ values are those in which $v_i = v_{max}$. In Figure 4.1, for example, the saturating $[S]$ values are those that are greater than about 4.5. To determine optimal temperature, a series of assays is performed at different temperatures. The activity for each assay is estimated and plotted against temperature. The optimal temperature corresponds to the highest activity. Optimal pH is determined similarly, except that the assay is performed at different pH values.

Data from assays that meet the above conditions are then used to calculate the enzyme's activity and kinetic constants using equations given in sections 4.3 and 4.4.

C. ASSAYS THAT MEASURE SUBSTRATE CONCENTRATION USING THE ENZYME AS A REAGENT

In region **B** of Figure 4.1, v_i varies linearly with $[S]$. When the rate of a reaction is dependent on the concentration of one reactant, the reaction has **first-order kinetics**. Region **B**, therefore, has first-order kinetics. $[S] \ll K_m$, and the $[S]$ in the denominator of equation 88 is small relative to K_m and can be ignored in the equation. In quantitative terms,

$$K_m + [S] \approx K_m$$

and equation 88 becomes

$$v_i = \frac{v_{max}}{K_m}[S] = k[S] \tag{90}$$

where $k = (v_{max}/K_m)$ is the first-order rate constant, and $[S]$ is the substrate concentration at the time v_i is measured. Equation 90 implies that when substrate concentration is limiting (i.e., $[S] \ll K_m$), the rate of an enzyme reaction is linearly proportional to the substrate concentration; the proportionality constant being v_{max}/K_m. This relationship forms the basis for setting up assays to measure the concentrations of substrates, using enzymes as reagents. The assays require the following conditions:

(a) The substrate concentration should be limiting (i.e., $[S] \ll K_m$) so that first-order kinetics are established.

(b) The enzyme concentration used should give a linear product formation or substrate consumption over a time period sufficiently long to permit making accurate measurements.

Data from assays that meet these conditions are then used to calculate the substrate's concentration, as explained below.

(1) Calculating $[S_0]$ based on kinetic data

The concentration to be determined corresponds to $[S_0]$, the substrate's concentration in the assay medium at $t = 0$. An equation that relates $[S_0]$ to the $[S]$ measured at a later time (t) during the kinetic phase of the enzyme reaction is derived by using equation 90 and the fact that v_i is the initial rate of decrease in the substrate's concentration as time changes:

$$v_i = \frac{d[S]}{dt} = k[S] \tag{91}$$

$$\therefore \; -\frac{d[S]}{[S]} = kdt \tag{92}$$

The following equations are obtained when equation 92 is integrated over the limits: $[S_0]$ to $[S]$ (for the left side) and 0 to t (for the right side). See Appendix A.5 for details.

$$[S] = [S_0]e^{-kt} \tag{93}$$

$$\therefore \; [S_0] = \frac{[S]}{e^{-kt}} \tag{94}$$

Equations 93 and 94 are the general expressions for a first-order reaction and are used to calculate $[S_0]$, if k is known ($k = v_{max}/K_m$). Alternatively, by taking readings at two times, t_1 and t_2, during the reaction, the change in substrate concentration (ΔS) during this interval can be used to calculate $[S_0]$ according to equation 95, below. See Appendix A.5 for details.

$$[S_0] = \frac{\Delta[S]}{(e^{-kt_1} - e^{-kt_2})} \tag{95}$$

(2) Determining [S_0] based on end-point assay

With this technique, the reaction is allowed to proceed to equilibrium so that practically all the substrate molecules have been converted to product before a reading is taken. Standards of the pure product are used to generate a standard curve from which the [S_0] of samples are read. The enzyme concentration in the assay should be sufficient for a complete conversion of the substrate within a reasonable assay time. If the equilibrium does not favor completion, the reaction can be coupled to a secondary reaction capable of displacing the equilibrium in favor of completion.

4.3. Enzyme Activity and Units

The activity of an enzyme is defined with respect to the amount of substrate catalyzed under a given set of assay conditions. It is calculated using raw data obtained from the type of assays described in section 4.2-B and the equations derived below.

A. ACTIVITY

This is defined as the amount of substrate converted or product liberated per unit time under a defined set of assay conditions. Activity is calculated using equation 96 or 97, below:

$$\text{activity} = \frac{\text{amount of } S \text{ converted}}{t} \tag{96}$$

or

$$\text{activity} = \frac{\text{amount of } P \text{ liberated}}{t} \tag{97}$$

Examples: See problems 35a and 37b in section 4.5.

B. SPECIFIC ACTIVITY (sp act.)

Specific activity is the amount of substrate converted or product liberated per unit time per milligram of protein under a defined set of assay conditions. **Specific activity is calculated using equation 98, 99, or 102, below:**

$$\text{sp act.} = \frac{\text{amount of } S \text{ converted}}{t \times \text{mg of protein}} \tag{98}$$

or

$$\text{sp act.} = \frac{\text{amount of } P \text{ converted}}{t \times \text{mg of protein}} \tag{99}$$

The "mg of protein" in the above equations refers to the total milligrams of all proteins present in the sample, not just in the enzyme. Decreasing the amount of contaminating proteins during purification decreases the denominator of equation 98, resulting in an increase in specific activity. Therefore, as the purity of an enzyme preparation increases, the specific activity of the enzyme also increases.

Examples: See problems 32a, 33, 35, 36, and 37 in section 4.5.

C. ENZYME ACTIVITY UNITS

An activity unit allows one to estimate the quantity of enzyme that possesses a given activity. In general, one unit of activity for an enzyme is the amount of the enzyme that converts a given amount of substrate per unit time under optimal assay conditions.

(1) International Unit (IU)

This widely used unit is defined as follows: One **IU** of an enzyme is the amount that catalyzes the conversion of 1 μmol of substrate per minute under a defined set of assay conditions. When expressed mathematically,

$$1 \text{ IU} = 1 \ \mu\text{mol/min} \tag{100}$$

The concentration of the enzyme can then be expressed as IU/unit volume of sample or IU/unit weight of proteins in the sample. To convert enzyme activity from other units to IU, first convert the units of S or P to μmol and the time to minutes, and seconds. Use equation 101 or 102, below, to calculate the activity in IU.

$$\text{activity (IU)} = \frac{\mu\text{mol of } S \text{ or } P \text{ catalyzed}}{\text{min}} \tag{101}$$

and

$$\text{sp act. (IU)} = \frac{\mu\text{mol of } S \text{ or } P \text{ catalyzed}}{\text{min} \times \text{mg of protein}} \tag{102}$$

Examples: See problems 36 and 37b in section 4.5.

(2) Katal (kat)

This unit is defined as the amount of enzyme that converts 1 mole of substrate per second. Thus,

$$1 \text{ kat} = 1 \text{ mol/s} = 6 \times 10^7 \ \mu\text{mol/min} = 6.0 \times 10^7 \text{ IU}$$

$$\therefore \ 1 \text{ IU} = 1/60 \ \mu\text{kat}.$$

4.4. Basic Kinetic Constants: K_{cat}, v_{max}, and K_m

Detailed discussions on enzyme kinetics can be found in references *6*, *9*, and *10*. Only the kinetic constants, K_{cat} (turnover number), v_{max}, and K_m, are defined and briefly discussed below.

A. TURNOVER NUMBER (K_{cat})

Turnover number or catalytic constant (K_{cat}), is the maximum number of moles of substrate converted to product per second per mole of active site on the enzyme. If there is one active site per enzyme molecule, then turnover number is the number of substrate molecules converted by one molecule of the enzyme per second. For example, a turnover number of 500 s^{-1} means that each enzyme molecule converts 500 molecules of substrate to product each second when substrate concentration is saturating. From the above definition,

$$\text{turnover no.} = \frac{v_{max}\,(\text{mol/s})}{[\text{enzyme}]\,(\text{mol})} = K_{cat} \qquad (103)$$

Rearrangement yields equation 104, which is used to calculate the turnover number of an enzyme.

$$\text{turnover no.} = \frac{\text{mol of } S \text{ catalyzed}}{\text{s} \times \text{mol of enzyme}} \qquad (104)$$

Example: See problem 35c in section 4.5.

B. MAXIMAL VELOCITY (v_{max})

This is the maximal velocity attained by an enzyme reaction when substrate concentration is saturating. At v_{max}, all the enzyme molecules are complexed with substrate molecules, and as soon as the bound substrate is converted and released, another one moves in for another round of turnover. v_{max} is a constant only when substrate concentration is saturating; its calculation is explained below.

C. MICHAELIS CONSTANT (K_m)

The K_m for an enzyme is the substrate concentration at which the rate of the enzyme's reaction is one-half of v_{max}. It is a constant characteristic of a given enzyme. The lower the K_m, the higher the affinity of the enzyme to the substrate, and when alternative substrates exist, the one with the lowest K_m fits best into the enzyme's active site. Knowledge of K_m facilitates (1) estimating the in vivo or in vitro optimum substrate concentration, (2) establishing how an enzyme is regulated, and (3) identifying differences in an enzyme isolated from different sources (*6*).

v_{max} and K_m are readily obtained from a plot of v_i versus $[S_0]$ (see Figure 4.1). More accurate values are obtained using a Lineweaver-Burk double reciprocal plot (Figure 4.3). The latter is a plot of equation 105, below, which is the inverse of equation 88:

$$\frac{1}{v_i} = \left(\frac{K_m}{v_{max}} \times \frac{1}{[S_0]} + \frac{1}{v_{max}}\right) \tag{105}$$

This being a linear equation means that

$$y \text{ intercept} = \frac{1}{v_{max}}$$

$$\therefore\ v_{max} = \frac{1}{y \text{ intercept}} \tag{106}$$

$$x \text{ intercept} = -\frac{1}{K_m}$$

$$\therefore\ K_m = -\frac{1}{x \text{ intercept}} \tag{107}$$

and

$$\text{slope} = \frac{K_m}{v_{max}} \tag{108}$$

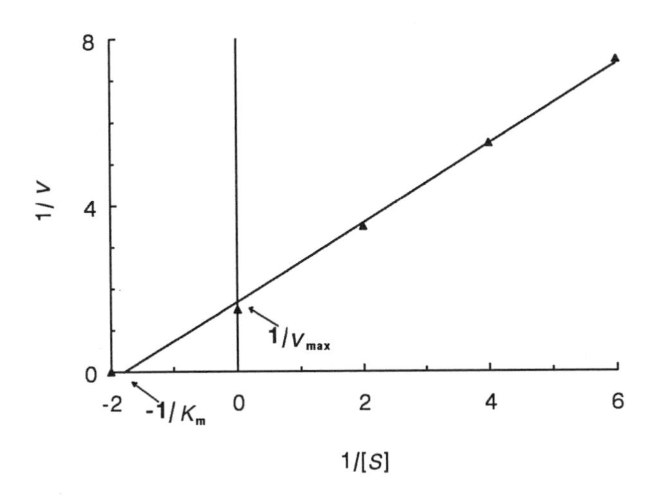

Figure 4.3. A Lineweaver-Burk double reciprocal plot (equation 105).

4.5. Practical Examples

PROBLEM 35

A 0.10 mL aliquot of a pure ß-galactosidase (ß-gal) solution (1.0 mg protein/mL) hydrolyzed 0.10 mmol of *o*-nitrophenyl galactoside (ONPG) in 5.0 min. Calculate (a) the activity, (b) the specific activity, and (c) the turnover number of the ß-gal. Assume a mol wt of 116 kD for the enzyme and one active site per molecule.

SOLUTION:

(a) Convert the amount of ß-gal hydrolyzed to μmoles, and the time (t) to minutes.

$$\text{amount of ß-gal hydrolyzed} = 0.10 \text{ mmol} \times 1000 \text{ (μmol/mmol)}$$

$$= 100 \text{ μmol}$$

$$t = 5.0 \text{ min}$$

Using equation 101,

$$\text{activity (IU)} = \frac{\text{μmol of ONPG hydrolyzed}}{\text{min}}$$

$$= \frac{100 \text{ μmol}}{5.0 \text{ min}} = 20 \text{ μmol/min} = 20 \text{ IU}$$

where 1 IU = μmol/min

(b) Total protein in assay = 1.0 mg/mL × 0.10 mL = 0.10 mg. From equation 102,

$$\text{sp act.} = \frac{\text{μmol of ONPG hydrolyzed}}{\text{min} \times \text{mg of protein}}$$

$$= \frac{100 \text{ μmol}}{5.0 \text{ min} \times 0.10 \text{ mg of protein}}$$

$$= 200 \text{ μmol min}^{-1} \text{ mg}^{-1} \text{ of protein}$$

$$= 2.0 \times 10^2 \text{ IU/mg of protein}$$

(c) Using equation 1,

$$\text{mol of enzyme in assay medium} = \frac{(0.00010 \text{ g})}{116{,}000 \text{ g/mol}}$$

$$= 8.6 \times 10^{-10} \text{ mol}$$

mol of ONPG hydrolyzed

$$= \frac{0.10 \text{ mmol}}{1000 \text{ mmol/mol}}$$

$$= 1.0 \times 10^{-4} \text{ mol}$$

Using equation 104,

$$K_{cat} \text{ turnover number} = \frac{\text{mol of ONPG hydrolyzed}}{\text{s} \times \text{mol of enzyme}}$$

$$= \frac{1.0 \times 10^{-4} \text{ mol}}{5.0 \text{ min} \times 60 \text{ (s/min)} \times 8.6 \times 10^{-10} \text{ mol}}$$

$$= 390 \text{ s}^{-1}$$

PROBLEM 36

A vial of lyophilized hexokinase (total protein, 100 mg; sp act. 200 IU/mg of protein) is reconstituted with 10.0 mL of diluent. (a) Calculate the units/mL of hexokinase in the solution. (b) What volume of this solution will yield 50 IU of hexokinase per mL of assay medium? (c) Repeat the calculations for a vial that contains 20,000 IU of total activity and is reconstituted with 5 mL of diluent.

SOLUTION:

(a)

$$\text{Units/mL} = \frac{\text{total activity}}{\text{total volume}}$$

$$= \frac{200 \text{ IU/mg of protein} \times 100 \text{ mg of protein}}{10 \text{ mL}}$$

$$= 2{,}000 \text{ IU/mL}$$

(b) Let V_1 be the volume of hexokinase solution needed to prepare the 1 mL (V_2) solution; C_1 and C_2 are the hexokinase concentrations in the stock and final solutions, respectively. By using equation 39,

$$V_1 = \frac{C_2 V_2}{C_1} = \frac{50\ \mu\text{mol min}^{-1}\ \text{mL}^{-1} \times 1\ \text{mL}}{2{,}000\ \mu\text{mol min}^{-1}\ \text{mL}^{-1}}$$

$$= 0.02\ \text{mL or } 20\ \mu\text{L}$$

∴ Twenty μL of the hexokinase solution is added to 0.980 mL of assay medium to yield 50 IU/mL.

(c) Calculate the hexokinase activity/mL of solution

$$\text{units/mL} = \frac{\text{total activity}}{\text{total volume}}$$

$$= \frac{20{,}000\ \text{IU}}{5.0\ \text{mL}}$$

$$= 4{,}000\ \text{IU/mL}$$

Following the example in (b),

$$V_1 = \frac{C_2 V_2}{C_1}$$

$$= \frac{50\ \text{IU/mL} \times 1\ \text{mL}}{4{,}000\ \text{IU/mL}}$$

$$= 0.01\ \text{mL or } 10\ \mu\text{L}$$

where C and V are defined as in (b), above.

Therefore, 10 μL of the hexokinase solution is added to 0.990 mL of the assay medium to yield 50 IU/mL.

PROBLEM 37

Alcohol dehydrogenase in a cell extract was assayed by adding 0.10 mL of the undiluted or diluted extract to 1.90 mL of the assay medium containing ethanol (ETOH) and NAD^+ in a 1 cm cuvette. The absorbance at 340 nm was monitored for 10 min and the data obtained are shown in Table 4.1, below. (a) Which data should be used to calculate the activity of alcohol dehydrogenase and why? (b) Calculate the total activity of alcohol dehydrogenase in 20.0 mL of the extract. The $E_{1\text{cm}}^{1\text{M}}$ of NADH at 340 nm is 6220.

TABLE 4.1. Data for Problem 37

	Time (min)						
OD	**0.0**	**2.0**	**4.0**	**6.0**	**8.0**	**10.0**	**DF**
(1)	0	0.75	0.97	1.04	1.06	1.09	1
(2)	0	0.40	0.70	0.90	0.97	1.04	3
(3)	0	0.12	0.24	0.36	0.48	0.60	10
(4)	0	0.06	0.12	0.18	0.24	0.30	20

SOLUTION:

(a) To determine which set of data should be used to calculate the activity, plots of $[P]$ versus t (Figure 4.4) and v versus t (Figure 4.5) are made. Data sets (1) and (2) should not be used to calculate the activity because the $[P]$ versus t plot is not linear and the v versus t plot is not constant for either set. Zero-order kinetics is required for assays to measure activity, in which case, v must be constant and $[P]$ must increase linearly with time (see section 4.2.B). These conditions are satisfied by data sets (3) and (4) (Figures 4.4 and 4.5); hence, either set can be used to calculate the activity. However, the reaction for data set (4) is somewhat too slow. Data set (3) is more appropriate for calculating the activity.

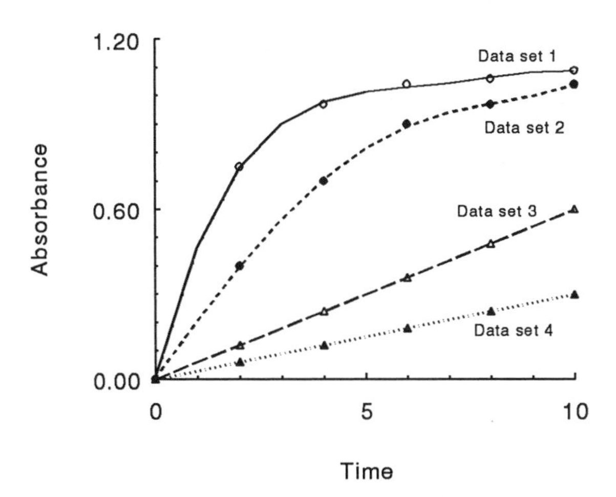

Figure 4.4. Plots of product concentration $[P]$ versus time for the data sets given in problem 37. $[P]$ is approximated by OD.

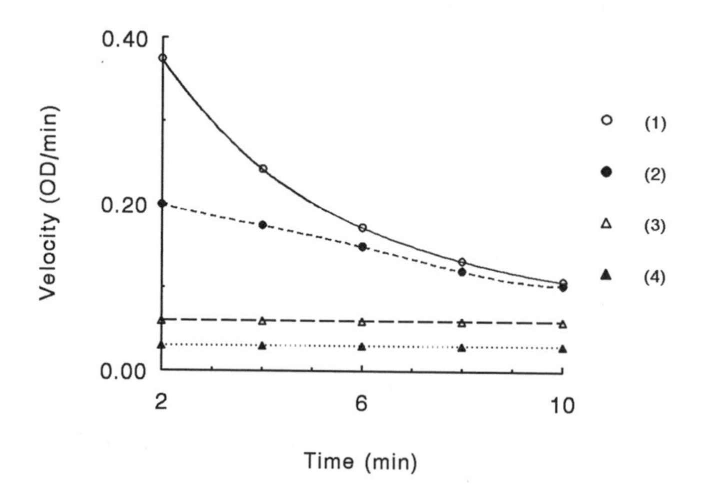

Figure 4.5. Plots of velocity versus time for the data sets given in problem 37. Velocity is approximated by OD/min.

(b) Any of the intervals within the linear phase of the reaction can be used for the calculations. For the interval between 0 and 8 min,

$$t = (8.0 - 0.0)\ \text{min} = 8.0\ \text{min};\ \text{OD} = (0.48 - 0) = 0.48.$$

First, using equation 84, calculate the amount of NADH formed:

$$[\text{NADH}] = \frac{A}{E\ell}$$

$$= \frac{0.48}{6220\ \text{M}^{-1}\ \text{cm}^{-1} \times 1\ \text{cm}}$$

$$= 7.72 \times 10^{-5}\ \text{M}$$

mole of NADH formed in 2-mL assay medium

$$= 7.72 \times 10^{-5}\ (\text{mol/L}) \times \frac{2.0\ \text{mL}}{1000\ (\text{mL/L})}$$

$$= 1.54 \times 10^{-7}\ \text{mol}$$

$$= 0.154\ \mu\text{mol}$$

Second, calculate the activity/mL of the extract using equation 97, and taking into account the dilution factor 10 and the fact that only 0.10 mL of the 20.0 mL extract was assayed.

$$\text{activity/mL} = \frac{0.154\ \mu\text{mol}}{10.0\ \text{min}} \times \frac{1}{0.10\ \text{mL}} \times 10$$

$$= 1.54\ \mu\text{mol min}^{-1}\ \text{mL}^{-1}$$

$$\text{Total activity} = \text{activity/mL} \times \text{total volume}$$

$$= 1.54\ \mu\text{mol min}^{-1}\ \text{mL}^{-1} \times 20.0\ \text{mL}$$

$$= 31\ \mu\text{mol/min} = 31\ \text{IU}$$

where 1 IU = 1 μmol/min.

Chapter 5

RADIOACTIVITY AND RELATED CALCULATIONS

5.1. Introduction

The isotopes of an element contain the same number of protons but different numbers of neutrons in the atomic nucleus. However, they have identical chemical properties because the number of protons is what determines the chemical properties of the element. Whether or not an isotope is stable depends on the relative numbers of protons and neutrons in its nucleus. When an isotope is unstable, it disintegrates to another isotope of the same or different element by emitting α, ß, γ, or other particles. Such an isotope is termed radioactive, and those that emit ß or γ particles are widely used in biological work.

A ß particle has a continuous range of energy values characteristic of the emitting isotope, and this is exploited in liquid scintillation to discriminate between different isotopes in a sample. A γ ray is a photon and has discrete energy values, in contrast to the continuous energy spectrum of the ß particle. A given isotope emits γ photons of one or more discrete energy values characteristic of the isotope. This is also exploited to discriminate between different γ-emitting isotopes in a sample. To measure the amount of radioactivity, either liquid scintillation or the Geiger-Müller counting technique is used to measure the rate of particle emission. The underlying principles for these techniques are briefly discussed in Appendix A.6.

Applications of radioactivity to studies in the life sciences consist of labeling biomolecules with atoms of a radioactive element such as ^{3}H, ^{14}C, ^{32}P, or ^{35}S, and using the labeled material to assay various cellular and biochemical functions. The quantity of radioactivity used is critical to the success of the assay, and the quantity recovered enables the experimenter to transform raw data to desired results. Basic calculations for experiments involving radioactivity are presented in this chapter.

5.2. Calculations Involving Radioactivity

Basic calculations involving radioactivity are performed on the basis of the quantitative descriptions of radioactive decay and the definitions of radioactivity units and specific radioactivity. These are discussed below.

A. QUANTITATIVE DESCRIPTION OF RADIOACTIVE DECAY

For a given radioactive substance, the number of decaying atoms (dN) per unit time interval (dt) is proportional to the initial number of radioactive atoms (N_0). When expressed mathematically,

$$-\frac{dN}{dt} = \lambda N_0 \qquad (109)$$

where λ is the proportionality constant. Both sides of equation 109 are then integrated from N_0 to N_t (left side), and from 0 to t (right side); N_0 and N_t being the number of radioactive atoms present at $t = 0$, and at a later time, t, respectively:

$$-\int_{N_0}^{N_t} \frac{dN}{N_0} = \int_0^t \lambda dt \qquad (110)$$

$$\ln \frac{N_0}{N_t} = \lambda t \qquad (111)$$

Taking the antilog,

$$N_t = N_0 e^{-\lambda t} \qquad (112)$$

Because the value of λ is not always available, equation 112 is modified so that a more readily available constant, $t_{½}$ **(half life)** can be used to calculate N_t. The $t_{½}$ is defined as the time it takes one-half of a given radioactive material to decay. If $N_0 =$ 1, then at $t_{½}$, $N_t = 0.5$. Substituting these into equation 112,

$$0.5 = e^{-\lambda t_{½}}$$

$$\ln 0.5 = -\lambda t_{½} \ln e = -\lambda t_{½}$$

$$\therefore \lambda = \frac{-\ln 0.5}{t_{½}} = \frac{0.693}{t_{½}}$$

Substituting this expression for in equation 112 yields

$$N_t = N_0 e^{-0.693t/t_{½}} \tag{113}$$

In practical terms, N_t is the amount of radioactivity remaining in a sample after a given time has elapsed, and is calculated using equation 113.

Example: See problem 40 in section 5.3.

B. UNITS OF RADIOACTIVITY

Common units of radioactivity are listed in Table 5.1 along with their equivalents in other units. Many calculations relating to radioactivity can be done based on these definitions.

Examples: See problems 38 and 39 in section 5.3.

C. SPECIFIC RADIOACTIVITY AND THE CALCULATION OF CONCENTRATIONS

In a labeling process, atoms of a specific element in a molecule are randomly replaced by radioactive atoms of the same element. Replacing too many atoms can adversely affect the function of the molecule because of high radiation effects; replacing too few can affect the sensitivity of the assay because the radiation signal would be too low for accurate detection. In practice, the proper amount is determined experimentally for particular molecules. To enable the experimenter to determine what quantity of the labeled material contains a given count rate and vice versa, the radioactivity is expressed per unit amount of the labeled material. **The quantity of radioactivity per unit amount of labeled material is termed specific radioactivity or, simply, specific activity.** Examples of specific radioactivity units include μCi/mg, mCi/μmol, cpm/mol, μCi/mL, and Bq/mg. To calculate specific activity, the following equation is used:

$$\text{sp radioactivity} = \frac{\text{radioactivity of a sample}}{\text{amount of the sample}} \tag{114}$$

Specific radioactivity can be used to calculate the amount of a radioactive substance needed to produce a given count rate, the concentration of a biological receptor that has been labeled with a radioactive ligand, or the intracellular concentration of solutes. These applications are best illustrated by specific **examples** such as problems 39, 40, 41, 42, and 43.

TABLE 5.1. Units of Radioactivity and Their Equivalents

Unit	Definition	Equivalent
Curie (Ci)	The amount of a radioactive substance decaying at a rate of 3.7×10^{10} disintegrations per second (dps).	1 Ci $= 3.7 \times 10^{10}$ dps $= 2.22 \times 10^{12}$ dpm
Microcurie (μCi)	One-millionth of a curie.	1 μCi $= 2.22 \times 10^{6}$ dpm
Disintegrations per minute (dpm)	The number of radioactive atoms disintegrating per minute.	
Counts per minute (cpm)	The number of disintegrations detected per minute. If the counting device is 100% efficient, then cpm is equal to dpm.	cpm = dpm × counting efficiency
Becquerel (Bq)	The SI unit of radioactivity defined as the quantity of a radioactive substance decaying at a rate of 1 dps.	1 Bq = 1 dps = 60 dpm $= 2.7 \times 10^{-11}$ Ci 1 Ci $= 3.7 \times 10^{10}$ Bq 1 mCi $= 3.7 \times 10^{7}$ Bq 1 μCi $= 3.7 \times 10^{4}$ Bq

5.3. Practical Examples

PROBLEM 38

Convert 1.0×10^{7} dpm to (a) μCi and (b) cpm. Assume a counting efficiency of 80%.

SOLUTION:

(a) From Table 5.1, 2.22×10^{6} dpm = 1 μCi

$\therefore$ the number of μCi in 1.0×10^{7} dpm

$$= \frac{1\ \mu Ci}{2.22 \times 10^6\ dpm} \times 1.0 \times 10^7\ dpm$$

$$= 4.5 \times \mu Ci$$

(b) From Table 5.1, cpm = dpm × counting efficiency

$$= 1.0 \times 10^7\ dpm \times \frac{80\ cpm}{100\ dpm} = 8.0 \times 10^6\ cpm$$

PROBLEM 39

(a) Starting with solid ^{14}C-inulin (sp act. 100.0 μCi/mg), prepare a 10 mL solution containing 500.0 μCi of radioactivity. (b) What volume of this solution will produce a count rate of 1.0×10^6 cpm in a 2.0 mL assay medium? (c) If 0.2 mL of the assay mixture is counted, what cpm is expected? Assume an 80% counting efficiency.

SOLUTION:

(a) The total μCi in the 10 mL solution = 500 μCi. From the given specific activity, 100.0 μCi is contained in 1 mg of the ^{14}C-inulin.

∴ 500.0 μCi is contained in

$$\frac{1\ mg}{100.0\ \mu Ci} \times 500.0\ \mu Ci = 5.0\ mg$$

To prepare the solution, weigh 5.0 mg of the ^{14}C-inulin and dissolve it such that the final volume is 10.0 mL. This solution contains a total of 500.0 μCi of radioactivity.

(b) Convert the count rate in the solution to cpm before calculating the required volume. From Table 5.1, 1 μCi contains 2.22×10^6 dpm.

$$\therefore\ 500.0\ \mu Ci \text{ contains } \frac{2.22 \times 10^6\ dpm}{1\ \ Ci} \times 500.0\ \mu Ci$$

$$= 1.11 \times 10^9\ dpm$$

$$= 1.11 \times 10^9 \times 80\%/100\% = 8.88 \times 10^8\ cpm$$

This 8.88×10^8 cpm is contained in 10 mL.

∴ The 1.0×10^6 cpm is contained in

$$\frac{10\ mL}{8.88 \times 10^8\ cpm} \times 1.0 \times 10^6\ cpm = 0.011\ mL \text{ or } 11\ \mu L$$

A solution volume of 11.3 μL contains the required 1.0×10^6 cpm in the assay medium.

(c) 2.0 mL of assay medium contains 1.0×10^6 cpm.

$\therefore$ The 0.2 mL contains

$$\frac{1.0 \times 10^6 \text{ cpm}}{2.0 \text{ mL}} \times 0.2 \text{ mL} = 1.0 \times 10^5 \text{ cpm}$$

PROBLEM 40

A purchased ^{32}P-labeled sample (4.5×10^7 cpm/pmol) could not be used until 6 days after it was prepared. Calculate the specific activity remaining in the sample. The $t_{½}$ of ^{32}P is 14.3 days.

SOLUTION:

Let the radioactivity remaining be N_t. Then, using equation 113,

$$N_t = N_0 e^{-0.693t/t_{½}}$$

$$= 4.5 \times 10^7 \text{ cpm/pmol} \times e^{(-0.693 \times 6 \text{ days})/14.3 \text{ days}}$$

$$= 3.4 \times 10^7 \text{ cpm/pmol}$$

Note: The units for t and $t_{½}$ must be the same.

PROBLEM 41

In a binding assay, labeled hormone (5.0×10^5 cpm/mg) was added to 1.0 mL of the assay medium containing the hormone's receptor. A 0.1 mL aliquot was then processed and counted for bound hormone. If the count rate was 1.0×10^4 cpm, calculate the total amount of hormone bound in the 1.0 mL of assay medium. The receptor has 1 binding site per molecule.

SOLUTION:

First, calculate the total bound cpm in the 1.0 mL of assay medium:

$$1.0 \times 10^4 \text{ cpm/0.1 mL} = 1.0 \times 10^5 \text{ cpm/mL}$$

Second, calculate the milligrams of hormone containing the 1.0×10^5 cpm. From the given specific activity, 5.0×10^5 cpm is contained in 1 mg of hormone.

$\therefore$ 1.0×10^5 cpm is contained in

$$\frac{1 \text{ mg}}{5.0 \times 10^5 \text{ cpm}} \times 1.0 \times 10^5 \text{ cpm} = 0.20 \text{ mg}$$

$\therefore$ bound hormone $= 0.20$ mg

PROBLEM 42

In a Na^+ transport assay, a total of 1.0×10^6 cpm of $^{22}NaCl$ (sp act., 4.0×10^8 cpm/μg) was added to 1.0 mL of the assay medium containing 1.5 mM nonradioactive NaCl and bacterium vesicles. (a) Calculate the specific activity of Na^+ in the medium. (b) If vesicles in 0.1 mL of the medium accumulated 5.0×10^3 cpm of the $^{22}Na^+$, calculate the moles of Na^+ accumulated.

SOLUTION:

The amount of Na^+ contributed by the $^{22}NaCl$ (less than 50 nmol) is negligible compared to contribution from the nonradioactive NaCl.

(a) total amount of Na^+ $= 1.5\ (\text{mmol/L}) \times 0.001\ \text{L} = 1.5\ \mu\text{mol}$

$$\text{total radioactivity} = 1.0 \times 10^6\ \text{cpm}$$

$$\therefore\ \text{sp act.} = \frac{1.0 \times 10^6\ \text{cpm}}{1.5\ \text{mol}} = 6.7 \times 10^5\ \text{cpm}/\mu\text{mol}$$

(b) 6.67×10^5 cpm is contained in 1 μmol of Na^+ in the medium.

$\therefore$ 5.0×10^3 cpm is contained in

$$\frac{1\ \mu\text{mol Na}^+}{6.67 \times 10^5\ \text{cpm}} \times 5.0 \times 10^3\ \text{cpm} = 0.0075\ \mu\text{mol Na}^+$$

In the 0.1 mL aliquot, 0.0075 μmol Na^+ (or 7.3 nmol Na^+) were accumulated by vesicles.

PROBLEM 43

A 50-base, single-strand synthetic DNA was purified and dissolved in 0.5 mL of buffer. If a 1 to 100 dilution of it has an OD of 0.25, (a) calculate the total moles of DNA in the solution. (b) If 10 pmol of the DNA is needed in a 20 μL assay mixture, what volume of the DNA solution should be added? (c) The 10 pmol were labeled with ^{32}P, and 70% was recovered and redissolved in 100 μL buffer. If 2 μL have a count rate of 450,500 cpm, calculate the specific activity of the labeled DNA.

SOLUTION:

(a) Using equation 87,

$$\text{concn of DNA} = 1\text{ AU} \times \text{OD} \times \text{DF}$$

$$= 50\ \mu\text{g/mL} \times 0.655 \times 100 = 3275\ \mu\text{g/mL}$$

$$\text{Total DNA} = 3275\ \mu\text{g/mL} \times 0.5\text{ mL} = 1637.5\ \mu\text{g}$$

Molecular mass of DNA = 650 D/bp × 50 bases = 32,500 D. Using equation 3,

$$\text{mol DNA} = \frac{\text{wt}}{\text{mol wt}} = \frac{1.64 \times 10^{-3}\text{ g}}{32{,}500\text{ g/mol}}$$

$$= 5.05 \times 10^{-8}\text{ mol}$$

$$= 5.05 \times 10^{-8}\text{ mol} \times \frac{1\text{ pmol}}{1 \times 10^{-12}\text{ mol}}$$

$$= 5 \times 10^{4}\text{ pmol}$$

(b) The concentration of DNA can be expressed as follows:

$$\frac{50{,}200\text{ pmol}}{500\ \mu\text{L}} = 100\text{ pmol}/\mu\text{L}$$

$\therefore$ volume containing 10 pmol is

$$\frac{1\ \mu\text{L}}{100\text{ pmol}} \times 10\text{ pmol} = 0.1\ \mu\text{L}$$

This volume is too small to pipette accurately with a micropipettor. Therefore, dilute an aliquot of the sample to a final concentration of 2 pmol/µL. Five µL of this solution contains 10 pmol of DNA. Use equation 39 to calculate V_1, the volume of the stock DNA solution that will yield 2 pmol/µL (C_2) in a final volume of 1 mL (V_2).

$$V_1 = \frac{C_2 V_2}{C_1}$$

$$= \frac{2\text{ pmol}/\mu\text{L} \times 1\text{ mL}}{100\text{ pmol}/\mu\text{L}} = 20\ \mu\text{L}$$

Therefore, add 20 µL of the stock DNA solution to 980.8 µL of diluent to obtain a solution of 2 pmol/µL. Add 5 µL of this solution to the reaction mixture.

(c) Since the recovery was 70%,

Amount of labeled DNA recovered

$$= \frac{70}{100} \times 10 \text{ pmol} = 7 \text{ pmol}$$

$\therefore$ the amount of DNA/2-μL aliquot

$$= \frac{7 \text{ pmol}}{100 \text{ μL}} \times 2 \text{ μL} = 0.14 \text{ pmol}$$

Radioactivity/2-μL aliquot = 450,500 cpm. Using equation 113,

$$\text{sp radioactivity} = \frac{\text{radioactivity of sample}}{\text{amount of sample}}$$

$$= \frac{450{,}500 \text{ cpm}}{0.14 \text{ pmol}}$$

$$= 3 \times 10^{-6} \text{ cpm/pmol}$$

References

1. Bloomfield, M.M. (1987) *Chemistry and the Living Organism.* John Wiley & Sons, New York.

2. Segal, B.G. (1989) *Chemistry, Experiment and Theory*, 2nd ed. John Wiley & Sons, New York.

3. Galister, H. (1991) *pH Measurement: Fundamentals, Methods, Applications and Instrumentation.* VCH Publishers, Inc., New York.

4. Georgiou, C.D., and Webster, D.A. (1987) *Biochemistry* **26**, 6521–6526.

5. Efiok, B.J.S., and Webster, D.A. (1990) *Biochem. Biophys. Res. Commun.* **173**, 370–375.

6. Zubay, G. (1988) *Biochemistry*, 2nd ed. Macmillan Publishing Co., New York. See pp 259–281 for a concise treatment of enzyme kinetics; and pp 236–238 for the application of spectrophotometry to monitor double- to single-stranded transitions of DNA.

7. Bashford, C. (1986). "An Introduction to Spectrophotometry and Fluorescence Spectrometry," in *Spectrophotometry & Spectrofluorimetry: A Practical Approach,* Bashford, C.L., and Harris, D.A., eds. IRL Press, Washington, D.C.

8. Cooper, T.G. (1977) *The Tools of Biochemistry*, John Wiley and Sons, New York, pp 59–62.

9. Lehninger, A.L. (1993) *Principles of Biochemistry.* Worth Publishers, Inc., New York. (See chapter on enzyme kinetics.)

10. Segel, I.H. (1976) *Biochemical Calculations*. John Wiley and Sons, New York. (A good reference for advanced and detailed treatment of biochemical calculations.)

11. Good, N.E., et al. (1966) *Biochemistry* **5**, 467–478; Good, N.E., and Izawa, S. (1968) *Hydrogen Ion Buffers. Methods in Enzymology* **24** (Pt. B), 53–68. (Sources for pK_a values.)

12. Friedlander, G., and Orr, W.C. (1951) *Phys. Rev.* **84**, 484; Laslett, L.J. (1949) *Phys. Rev.* **76**, 858. (Sources for Isotope half-life.)

Appendix A

1. Derivation of the Ionization Constant of H_2O

Pure H_2O ionizes as follows:

$$H_2O \rightleftharpoons H^+ + OH^-$$

and the equilibrium constant is

$$K_{eq} = \frac{[H^+][OH^-]}{[H_2O]}$$

Rearrangement yields

$$K_{eq}[H_2O] = [H^+][OH^-]$$

The concentration of H_2O in aqueous solution is practically constant (about 55.6 M) because at 25°C, only about 10^{-7} M dissociates. Therefore, the term, $K_{eq}[H_2O]$ is a constant and is defined as K_w, the dissociation constant of H_2O.

$$\therefore K_w = [H^+][OH^-]$$

2. Derivation of the Henderson-Hasselbach Equation

A weak acid ionizes in H_2O as follows:

$$HA + H_2O \rightleftharpoons H_3O^+ + A^-$$

The ionization constant (K_i) is:

$$K_i = \frac{[H_3O^+][A^-]}{[H_2O][HA]}$$

Multiplying both sides of the equation by $[H_2O]$ yields:

$$K_i[H_2O] = \frac{[H_3O^+][A^-]}{[HA]}$$

$[H_2O]$ is large (55.6 M) and essentially constant (in pure H_2O at 25°C, only 10^{-7} M is ionized). Therefore, $K_i[H_2O]$ is a constant defined as K_a, the acid dissociation constant. Also, $[H_3O^+] = [H^+]$. Substituting these into the above equation yields:

$$K_a = \frac{[H^+][A^-]}{[HA]}$$

Taking the logarithm of the above equation:

$$\log K_a = \log \frac{[H^+][A^-]}{[HA]} = \log ([H^+][A^-]) - \log [HA]$$

$$= \log [H^+] + \log [A^-] - \log [HA]$$

$$-\log [H^+] = -\log K_a + \log [A^-] - \log [HA]$$

$$-\log [H^+] = -\log K_a + \log \frac{[A^-]}{[HA]}$$

By definition, pH $= -\log [H^+]$ and p$K_a = -\log K_a$. Substituting these into the above equation yields:

$$pH = pK_a + \log \frac{[A^-]}{[HA]}$$

This is the Henderson-Hasselbach equation, written for a weak acid. For a weak base (R-NH_2), the equation becomes:

$$pH = pK_a + \log \frac{[R\text{-}NH_2]}{[R\text{-}NH_3^+]}$$

3. Derivation of the Beer-Lambert Equation

Lambert's law states that when monochromatic incident light of intensity I_0 passes through a solution of a light-absorbing substance, the intensity (I) of the transmitted light decreases as an exponential function of the path length ℓ (Figure A.3-1).

From the above definition, $I = I_0\, 10^{-k\ell}$,

$$\therefore \frac{I}{I_0} = 10^{-k\ell}$$

Similarly, Beer's law states that I decreases as an exponential function of the concentration (C) of the light-absorbing substance. Thus,

$I = I_0\, 10^{-k'C}$,

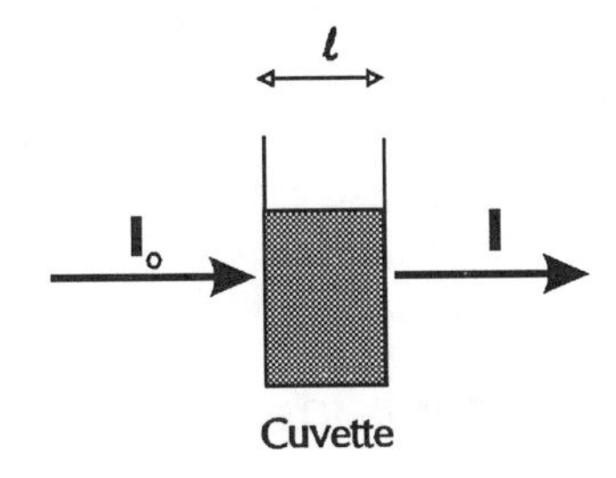

Figure A.3-1. Incident and transmitted light beams through a light-absorbing solution.

$$\therefore \frac{I}{I_0} = 10^{-k'C}$$

Combining the Lambert and Beer equations,

$$\frac{I}{I_0} = 10^{-E\ell C}$$

where E is a combined constant for k and k'.

Upon taking the logarithm,

$$\log \frac{I}{I_0} = \log 10^{-E\ell C} = -E\ell C$$

$$-\log = \frac{I}{I_0} E\ell C, \quad \text{or} \quad \log \frac{I_0}{I} = E\ell C$$

The term, $\log (I_0/I)$, is defined as absorbance (A),

$$\therefore A = E\ell C$$

This is the Beer-Lambert equation. For transmittance (T),

$$T = \frac{I}{I_0}$$

Also,

$$\%T = \frac{I}{I_0} \times 100\%$$

4. Spectrophotometers and Absorbance Measurement

A spectrophotometer is an instrument used to measure the optical absorbance or transmittance of a substance. It measures the incident and emergent light beams and calculates the absorbance or transmittance using these measured values (see Appendix A.3). The instrument can be classified as single beam or dual beam.

A. SINGLE-BEAM SPECTROPHOTOMETERS

These machines use a single light path for measurements. Their six essential components (Figure A.4-1) are as follows:

(1) **Light source:** A tungsten or Xenon arc lamp is commonly used for measurements in the visible region (400–900 nm), and a hydrogen lamp is used for measurements in the UV region (200–400 nm). UV-visible spectrophotometers have both types of lamp, and selection is either automatic or manual. The major requirement for these lamps is that they have stable and high-intensity output.

(2) **Monochromator:** A light beam from the lamp passes through a monochromator assembly where it is separated into a continuous spectrum of wavelengths by a prism or a diffraction grating. By adjusting the wavelength selector on the machine, a slice of the light spectrum centering about the wavelength of choice is isolated and passed through a slit (exit) of the monochromator assembly. (The monochromator does not select light of a single wavelength but a small spectrum of light of the same color having a maximum intensity at the chosen wavelength. In older machines, the monochromator is simply a filter that transmits light of the same color.)

(3) **Slit:** The incident beam exiting the monochromator passes through a slit that controls the wavelength purity as well as the intensity of the beam. The slit width may be fixed or adjustable (most simple spectrophotometers have fixed slit width). Decreasing the slit width increases the wavelength purity and, as such, the resolution or specificity of the measurement. However, this leads to a decrease in the intensity of the incident light and results in a decreased sensitivity. The reverse occurs when the slit width is increased. Choosing the right slit width, therefore, involves striking a balance between the desired resolution and sensitivity.

(4) **Sample cuvette:** From the slit, the incident beam passes through a sample cuvette positioned inside a sample housing. The housing is covered to prevent external light from reaching the photodetector and interfering with measurements. When more than one cuvette is used to measure a set of samples, the cuvettes should be optically identical or balanced so that errors arising from differences in their path length and light transmission properties cancel out. Quartz or silica cuvettes are used when measuring absorbance in the UV region

because they do not absorb light significantly either in the UV or the visible region. Glass cuvettes absorb strongly in the UV region and are therefore unsuitable, except for measurements in the visible region where they do not absorb significantly.

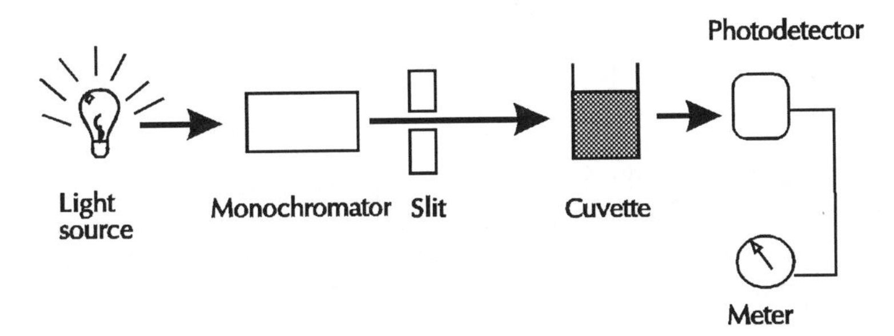

Figure A.4-1. Components of single-beam spectrophotometers. See text for explanation.

(5) **Light detector/display devices:** Light transmitted through the sample is measured and amplified by a photosensitive detector (photomultiplier or photocell/amplifier pair), and the signal is converted to absorbance or transmittance and displayed on a meter or a chart recorder.

B. DUAL-BEAM SPECTROPHOTOMETERS

Dual-beam spectrophotometers differ from single-beam machines in that they use two light paths for measurements: The incident light leaving the monochromator is split into a reference beam and a sample beam. The reference beam is passed through a cuvette containing a blank or reference solution, and the sample beam is passed through a cuvette containing a sample. Light emerging from each cuvette is detected by a photodetector, and the machine automatically subtracts the reference signal from the sample signal and uses the net value to calculate the sample's absorbance, transmittance, or concentration. The dual-beam technique compensates for errors due to fluctuations in the intensity of the lamp's output.

Some-dual beam spectrophotometers are simple machines, but others are sophisticated: They are capable of automatically scanning the UV and visible wavelength regions and simultaneously recording the absorbance, transmittance, or concentration of a sample. Additionally, they have optional components and internal microprocessors that facilitate fine-tuning the sensitivity and specificity of measurements, as well as storing and analyzing scanned spectra.

C. GENERAL PROCEDURE FOR ABSORBANCE MEASUREMENT

(1) Single-Beam Simple Spectrophotometers

(a) Turn on the machine and let it warm up for 15–20 minutes. If a UV wavelength will be used, also turn on the hydrogen lamp.

(b) Select the wavelength using the wavelength selector.

(c) Select the appropriate cuvette: quartz or silica for UV/visible, or glass for visible wavelengths.

(d) Calibrate the absorbance/transmittance scale as follows:

(i) With no cuvette in position, set the absorbance to ∞, or the transmittance to zero: When a cuvette is not inserted, the incident light is prevented from reaching the photo- detector. The signal across the photodetector represents the maximum absorbance or zero transmittance.

(ii) Fill the cuvette with an adequate volume of a blank solution and position it in the sample housing. With the housing door closed, set the absorbance to zero using the zero adjust, or the transmittance to 100%. The blank solution should best approximate the test solution, but lacks the assay's test color.

(e) Decant the blank solution, rinse the cuvette and fill it with the sample or test solution. Insert the cuvette as before and read the absorbance or % transmittance of the sample.

(2) Dual-Beam Spectrophotometer

Specific instructions for using these machines are best described in their operation manuals. The basics are similar to that described above except that:

(a) The blank solution should be in both the reference and the sample positions before setting the zero absorbance.

(b) When measuring the absorbance for samples, the blank should always be in its position.

(c) Some machines set the maximum absorbance automatically.

5. Derivation of the First-Order Rate Equation

Equation 90 showed that when $[S]$ is limiting in an enzyme assay, the velocity (v) is proportional to the substrate concentration ($[S]$):

$$v = k[S]$$

k being the proportionality constant. The velocity is also defined as the decrease in $[S]$ per unit change in time:

$$v = -\frac{d[S]}{dt}$$

Combining the two equations,

$$v = k[S] = -\frac{d[S]}{dt}$$

or

$$-\frac{d[S]}{[S]} = kdt$$

This equation is then integrated from $[S_0]$ to $[S]$ for the left side, and from 0 to t for the right side; $[S_0]$ and $[S]$ being the substrate concentrations at time $t = 0$ and a later time t, respectively.

$$-\int_{S_0}^{S(t)} \frac{d[S]}{[S]} = \int_0^t dt$$

$$-(\ln [S] - \ln [S_0]) = k(t-0)$$

$$\ln \frac{[S_0]}{[S]} = kt$$

Taking the antilog,

$$\frac{[S_0]}{[S]} = e^{kt}$$

$$\therefore \; [S_0] = [S]e^{kt} \text{ or } [S] = [S_0]e^{-kt}$$

The change in substrate concentration ($\Delta[S]$) within the time interval t_1 to t_2 can be related to $[S_0]$. Starting from the last equation,

$$\Delta[S] = [S_{t_1}] - [S_{t_2}] = [S_0]\mathrm{e}^{-kt_1} - [S_0]\mathrm{e}^{-kt_2}$$

$$\text{i.e.,} \quad \Delta[S] = [S_0](\mathrm{e}^{-kt_1} - e^{-kt_2})$$

or

$$[S_0] = \frac{\Delta[S]}{(e^{-kt_1} - e^{-kt_2})}$$

6. Liquid Scintillation and Geiger-Müller Techniques for Radioactivity Measurements

The general principle underlying these techniques for radioactivity measurement is as follows: When a ß particle collides with an atom, the atom absorbs the particle's energy and becomes either ionized or excited to a higher energy state. Liquid scintillation counting is based on detecting the fluorescence that accompanies the return of the excited atom to ground state, whereas the Geiger-Müller technique is based on detecting the ionized atom.

A. GEIGER-MÜLLER TECHNIQUE

This technique counts mostly ß particles. In general, a ß particle enters a gas-filled tube (Geiger tube) in which an electric field has been applied. Upon collision with an atom of the gas, the particle's energy is absorbed by the atom which then loses an electron and produces an ion pair—an electron that is attracted to the anode, and a positively charged ion of the gas that is attracted to the cathode. The electrodes detect and record these ions as electrical pulses, representing counts.

B. LIQUID SCINTILLATION TECHNIQUE

This is used to count both ß and γ particles, but the detector used for each type of particle differs.

(1) **ß Counting:** The overall strategy is to set up an assay medium in which its components will efficiently transfer the energy of a ß particle to fluorescence compounds that emit light of a wavelength suitable for detection by a photodetector.

In practice, the radioactive sample is mixed with a liquid scintillation "cocktail" that contains an excitable solvent (usually toluene) and one or more fluorescent compounds short-named "fluors." A ß particle emitted by a radioactive sample collides with and transfers some or all of its energy to a solvent molecule which becomes excited. The excited solvent molecule either transfers its energy to another solvent molecule, or phosphoresces and emits the excitation energy as light having a wavelength that is usually too short for detection by the instrument's phototube. This photon is absorbed by a primary fluor (F_1), which becomes excited, fluoresces, and emits a photon of longer wavelength. If its wavelength is sufficiently long, the photon is detected and counted by the phototube; otherwise, it is reabsorbed by a secondary fluor, which subsequently fluoresces and emits a photon with a longer wavelength for a subsequent detection. If the wavelength is still too short, additional fluors with suitable characteristics are added to the cocktail. A summary of the net reaction is as follows:

$$\text{solvent} + ß^- \rightarrow ß \text{ (lower energy)} + \text{solvent}^*$$

$$\text{solvent}^* + F_1 \rightarrow \text{solvent} + F_1^*$$

$$F_1^* \rightarrow F_1 + (hv)_1$$

$$\text{or } F_1^* + F_2 \rightarrow F_1 + F_2^*$$

$$F_2^* \rightarrow F_2 + (hv)_2$$

where (hv) is the photon emitted by the fluorescent compounds, which is detected by the phototube and reported as counts per minute.

(2) γ **Counting:** The liquid scintillation fluid described above for ß particles is unsuitable for counting γ photons. This is because a γ photon does not have mass; hence, it penetrates matter deeply and, therefore, requires a medium denser than a liquid to be absorbed efficiently. A fluorescent NaI crystal cell is the medium commonly used. The sample (usually a solid in a vial) is placed in the NaI cell. The γ photon leaving the sample penetrates and interacts with the NaI crystal producing ß particles which excite adjacent portions of the crystal and cause fluorescence. Light photons from the fluorescence are detected and counted by the phototube.

Appendix B

Dealing with Numbers

1. Handling Exact and Experimental Numbers

Data used for laboratory calculations consist of exact and/or measured numbers. Exact numbers are those whose values are exactly known. They include numbers that result from fundamental definition of quantities or from counting objects. For example, 1 L = 1000 mL; the 1000 is an exact number because its exact value is known. Exact numbers have an infinite number of **significant digits** or **significant figures**. If this were to be indicated numerically, the 1000 above would be written as 1,000.00000 . . . infinity. For convenience, the zeros to the right of the decimal point are omitted, but their presence must be assumed when deciding the number of significant figures in a value calculated from data containing both exact and measured values.

In contrast, numbers that are experimentally measured are not exact because of small errors or uncertainties associated with laboratory measurements. All laboratory instruments and techniques have limited accuracy, which in turn imparts small errors or uncertainty to experimental values measured with them. Scientists indicate these uncertainties by using an appropriate number of significant digits or significant figures in the measured values **(a significant digit is one whose value is reasonably reliable).** This is done by recording all digits that are known with certainty, and then adding an extra digit that has an uncertainty. To illustrate, if the light absorption of a sample is being measured using a spectrophotometer, and the meter indicates the reading shown in Figure B.1-1, then the first two digits of the result (0.52) are certain because the indicator points between the calibration marks 0.52 and 0.53. The third digit is uncertain because the space between the two numbers is not clearly marked. Its estimated value is 6, and unless specified otherwise, an estimated digit is assumed to have a range of ±1. Therefore, the result is recorded as 0.526, but its exact value lies between 0.525 and 0.527. Estimating the result to four decimal places such as 0.5258 does not improve its accuracy because the instrument is limited to only three decimal places.

2. Determining Significant Figures

In any number, the digits 1 through 9 are all significant. Thus the numbers 561 and 0.56123 have three and five significant figures, respectively.

The digit 0 may or may not be significant, depending on where it appears in a number.

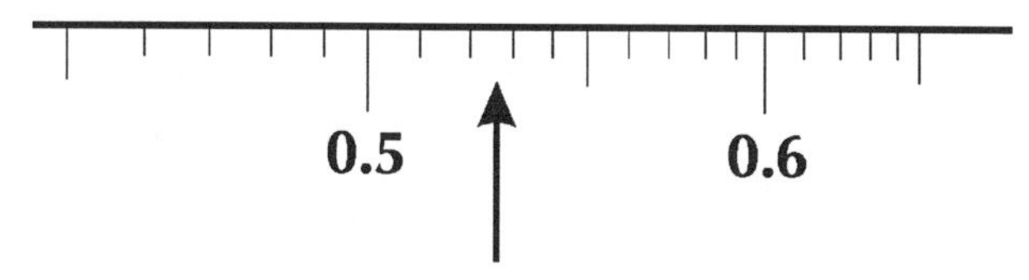

Figure B-1.1. Recording experimental data.

(a) **If the number is less than 1, all zeros to the right of the first nonzero digit are significant, but zeros to the left are not.** For example, the number 0.037700 has five significant figures because the two zeros that precede the first nonzero digit 3, are not significant, whereas the two zeros to the right are.

(b) **If the number is greater than 1, and there are digits on both sides of the decimal point, then all zeros are significant.** Example, in 5.0003 and 50.300, all the zeros are significant, hence, each has five significant figures.

(c) **If the number is greater than 1 and there are no digits to the right of the decimal point, then zeros that are located between two nonzero digits are significant. Zeros to the right of the last nonzero digit are not significant unless indicated otherwise.** For example, 70,029 has five significant figures because the two zeros between the digits 7 and 2 are significant. Each of the numbers 729, and 72,900 has three significant figures because the zeros in 72,900 are assumed not significant. If these zeros are significant, the number is expressed in exponential form, 7.2900×10^4, to indicate five significant figures.

3. Rounding Off Numbers

When experimental data or calculated results have more digits than are needed to indicate reliability, they are **rounded off** by eliminating the excess digits. The rules for rounding off are summarized below.

(a) If the first digit dropped is greater than a 5, increase the last retained digit by 1. For example, 3.3772 becomes 3.38 when rounded to two decimal places or three significant figures.

(b) If the first digit dropped is less than a 5, do not change the last retained digit. Using the above example, 3.3772 becomes 3.377 when rounded off to three decimal places or four significant figures.

(c) If the last digit dropped is a 5, do as follows: If the 5 is followed by digits greater than zero, increased the last retained digit by 1. For example, 3.88512 becomes 3.89 when rounded to two decimal places.

If the 5 is followed by a zero or no other digit, increase the last retained digit by 1 if that digit is odd, but leave it unchanged if it is even. For example, 3.885 or 3.88501 become 3.88 when rounded to three digits. In contrast, 3.875 or 3.8750 becomes 3.88 when rounded to three digits.

4. Significant Digits and Rounding in Calculated Values

Results calculated from experimentally derived data should retain an appropriate number of significant figures or decimal places to indicate the reliability of the calculated values. Here are the rules:

(a) Decide how many significant figures or decimal places the final answer should have [see rules (c), (d), and (e), below].

(b) If a result is to be used for further calculations, retain one or more significant digits than is necessary. If the result is final, round it to the number of significant figures estimated in step (a).

(c) If the calculations involve division and/or multiplication only, report the final result to as many significant figures as there are in the term with the least number of significant figures. For example,

$$\frac{37.0}{36.51 \times 0.08403} = 12.06 = 12.1$$

The 12.1 has been rounded to the same number of significant digits as 36.0; the factor with the least number of significant figures.

(d) If the calculations involve addition and/or subtraction, report the final result to as many decimal places as there are in the term with the least number of decimal places. For example,

$$12.046 + 9.2 - 0.345 = 20.901 = 20.9$$

The answer has been rounded to 20.9 because the term 9.2 has the least number of decimal places, which is one.

(e) If the calculations involve division/multiplication and/or addition/subtraction, follow the above rules, but perform the addition/subtraction first. In the example below, the addition operation in the numerator is performed first. Because its answer is to be used for a further calculation, it retains one more significant figure than would be required if it were the final answer. The division is performed last, and the final answer, 6.7, has been rounded to one decimal place in accordance with the above rule.

$$\frac{0.5 + 0.075}{0.075} = \frac{0.575}{0.075} = 7.6667 = 7.7$$

Appendix C

International System of Units (SI)

With a few exceptions, most of the units used in this book are SI (Systeme Internationale). It is a system of units adopted in 1960 by an international organization to bring worldwide uniformity to scientific measurements. Some of the fundamental units are shown in Table C-1, and the prefixes used to indicate their fractions and multiples are shown in Table C-2. For example, the prefixes for 1×10^{-3} and 1×10^{-6} of a unit are milli- (m) and micro- (μ), respectively. For the mass unit gram (g), 1 milligram (mg) = 1×10^{-3} g, and 1 μg = 1×10^{-6} g. Similarly, for the amount unit mole (mol), 1 mmol = 1×10^{-3} mol, and 1 μmol = 1×10^{-6} mol. See Quick Reference to Units (page vii) and Tables C-3 through C-6 below, for more examples.

Table C-1. Selected SI Units

Quantity	Unit	Symbol
Mass	kilogram	kg
Volume	liter	L
Amount of substance	mole	mol
Length	meter	m
Time	second	s
Electric current	ampere	A
Temperature	Kelvin	K

Table C-2. Prefixes for SI Units

Prefix	Symbol	Multiples and Submultiples
exa	E	10^{18}
peta	P	10^{15}
tera	T	10^{12}
giga	G	10^{9}
mega	M	10^{6}
kilo	k	10^{3}
deci	d	10^{-1}
centi	c	10^{-2}
milli	m	10^{-3}
micro	μ	10^{-6}
nano	n	10^{-9}
pico	p	10^{-12}
femto	f	10^{-15}
atto	a	10^{-18}

INSTRUCTIONS FOR TABLES C-3 throughC-6.

Definitions are: source unit, the unit to be converted; target unit, the unit into which the source unit is to be converted. To convert from one unit to another, select the **source unit** from the **boldfaced** set (leftmost column). Move in a straight line across the corresponding row to a factor directly under the *italicized target unit* (topmost row). Multiply the data by this factor. For example, to convert 0.5 **mg** to *μg*, the **mg** is selected from the boldfaced set in the leftmost column of units in Table C-3. Moving straight across the corresponding row, the conversion factor 10^3 is directly under the *italicized* μg in the top row of units. This means that:

$$1 \text{ mg} = 10^3 \ \mu\text{g}; \ \therefore \ 0.5 \text{ mg} = (0.5 \times 10^3) = 500 \ \mu\text{g}$$

If the source unit is selected from the *italicized* set instead, follow the above instructions except that the target unit is selected from the boldfaced set, and the data are divided by the factor. Using the above example, the factor obtained is 10^{-3}.

$$\therefore \ 0.5 \text{ mg} = 0.5 \div 10^{-3} = 500 \ \mu\text{g}.$$

Table C-3. Conversion Factors for Gram Units

	Target						
Source	*Kg*	*g*	*mg*	*μg*	*ng*	*pg*	*fg*
kg	**10^{0}**	10^{3}	10^{6}	10^{9}	10^{12}	10^{15}	10^{18}
g	10^{-3}	**10^{0}**	10^{3}	10^{6}	10^{9}	10^{12}	10^{15}
mg	10^{-6}	10^{-3}	**10^{0}**	10^{3}	10^{6}	10^{9}	10^{12}
μg	10^{-9}	10^{-6}	10^{-3}	**10^{0}**	10^{3}	10^{6}	10^{9}
ng	10^{-12}	10^{-9}	10^{-6}	10^{-3}	**10^{0}**	10^{3}	10^{6}
pg	10^{-15}	10^{-12}	10^{-9}	10^{-6}	10^{-3}	**10^{0}**	10^{3}
fg	10^{-18}	10^{-15}	10^{-12}	10^{-9}	10^{-6}	10^{-3}	**10^{0}**

Table C-4. Conversion Factors for Liter Units

	Target					
Source	*kL*	*L*	*dL*	*mL*	*μL*	*nL*
kL	**10^{0}**	10^{3}	10^{4}	10^{6}	10^{9}	10^{12}
L	10^{-3}	**10^{0}**	10^{1}	10^{3}	10^{6}	10^{9}
dL	10^{-4}	10^{-1}	**10^{0}**	10^{2}	10^{5}	10^{8}
mL	10^{-6}	10^{-3}	10^{-2}	**10^{0}**	10^{3}	10^{6}
μL	10^{-9}	10^{-6}	10^{-5}	10^{-3}	**10^{0}**	10^{3}
nL	10^{-12}	10^{-9}	10^{-8}	10^{-6}	10^{-3}	**10^{0}**

Table C-5. Conversion Factors for Meter Units[a]

	Target						
Source	*Km*	*m*	*cm*	*mm*	*μm*	*nm*	*Å*
Km	**10^{0}**	10^{3}	10^{5}	10^{6}	10^{9}	10^{12}	10^{13}
m	10^{-3}	**10^{0}**	10^{2}	10^{3}	10^{6}	10^{9}	10^{10}
cm	10^{-5}	10^{-2}	**10^{0}**	10^{1}	10^{4}	10^{7}	10^{8}
mm	10^{-6}	10^{-3}	10^{-1}	**10^{0}**	10^{3}	10^{6}	10^{7}
μm	10^{-9}	10^{-6}	10^{-4}	10^{-3}	**10^{0}**	10^{3}	10^{4}
nm	10^{-12}	10^{-9}	10^{-7}	10^{-6}	10^{-3}	**10^{0}**	10^{1}
Å[b]	10^{-13}	**10^{-10}**	**10^{-8}**	**10^{-7}**	**10^{-4}**	**10^{-1}**	**10^{0}**

[a] See instructions on page 168.
[b] Not an SI unit.

Table C-6. Conversion Factors for Mole-Related Units[a]

	Target					
Source	*mol*	*mmol*	*μmol*	*nmol*	*pmol*	*fmol*
mol	**10^{0}**	10^{3}	10^{6}	10^{9}	10^{12}	10^{15}
mmol	10^{-3}	**10^{0}**	10^{3}	10^{6}	10^{9}	10^{12}
μmol	10^{-6}	10^{-3}	**10^{0}**	10^{3}	10^{6}	10^{9}
nmol	10^{-9}	10^{-6}	10^{-3}	**10^{0}**	10^{3}	10^{6}
pmol	10^{-12}	10^{-9}	10^{-6}	10^{-3}	**10^{0}**	10^{3}
fmol	10^{-15}	10^{-12}	10^{-9}	10^{-6}	10^{-3}	**10^{0}**

[a] To use with molarity (M), Equivalence (equiv), or normality (N), substitute mol with M, equiv, or N.

Appendix D

Commercial Concentrated Acids and Bases

Table D-1. Concentrations of Commercial Concentrated Acids and Bases

Acid or Base	Mol Wt	% (w/w)	Sp Gr[a]	Approx. Molarity	Approx. Normality
Acetic acid	60.05	99.7	1.05	17.43	17.4
Ammonium hydroxide	35.05	28	0.90	14.8	14.8
Formic acid	46.03	97	1.22	25.7	25.7
Hydrochloric acid	36.46	37	1.20	12.18	12.1
Lactic acid	90.08	85	1.21	11.42	11.4
Nitric acid	63.01	70	1.40	15.55	15.5
Perchloric acid	100.46	70	1.66	11.57	11.6
Phosphoric acid	98	85	1.69	14.66	44.1
Sulfuric acid	98.1	95	1.84	17.82	35.6

[a] Specific gravity.

Table D-2. Volumes of Commercial Concentrated Acids and Bases Needed to Prepare Dilute Solutions

	mL Required to Prepare 1 L Solutions		
Acid or Base[a]	**0.1 N**	**0.5 N**	**1.0 N**
Acetic acid	5.7	28.7	57.4
Ammonium hydroxide	13.4	67.1	134.2
Formic acid	3.9	19.5	38.9
Hydrochloric acid	8.2	41.1	82.1
Lactic acid	8.7	43.8	87.6
Nitric acid	6.4	32.2	64.3
Perchloric acid	8.6	43.3	86.4
Phosphoric acid	2.3	11.4	22.7
Sulfuric acid	2.8	14.0	28.1

[a] See p. 191 for concentrations.

Appendix E

Table E-1. pK_a Values of Selected Weak Acids and Bases

Free Acid[a]	**Anhydrous Mol Wt**	**pK_{a_1}**	**pK_{a_2}**	**pK_{a_3}**
Acetic acid	60.05	4.76	—	—
Boric acid	61.80	9.23		
CAPS	221.32	10.40	—	—
Carbonic acid	62	6.10	10.25	—
Citric acid	192.12	3.09	4.75	5.40
Diethylamine	73.14	10.98	—	—
EDTA[b]	292.24	1.70	2.60	6.30
EPPS	252.33	8.0	—	—
Ethylamine	45.09	10.75	—	—
Glycine	75.07	2.40	9.6	—
Glycylglycine	132.12	3.10	8.4	—
HEPES	238.31	7.55	—	—
Histidine	155.16	1.82	6.0	9.17
Imidazole	68.08	7.00	—	—
MES	195.20	6.15	—	—
MOPS	209.26	7.2	—	—
MOPSO	225.27	6.9	—	—
PIPES	302.37	3.0	6.8	—
Phosphoric acid	98.00	1.96	6.8	12.32
Phthalic acid	166.13	2.90	5.4	—
Pyridine	79.10	5.20	—	—
Succinic acid	118.09	4.19	5.55	—
TAPS	243.28	8.4	—	—
TES	229.25	7.5	—	—
TRICINE	180.18	8.15	—	—
Triethylamine	101.19	10.70	—	—
TRIS	121.14	8.10	—	—

[a] CAPS, 3-cyclohexylamino-1-propane sulfonic acid; EDTA, ethylenediaminetetraacetic acid; EPPS, 4-(2-hydroxyethyl)-1-piperazine propanesulfonic acid; HEPES, 4-(2-hydroxyethyl)-1-piperazine ethanesulfonic acid; MES, 4-morpholine ethanesulfonic acid; MOPS, 4-morpholine propanesulfonic acid; MOPSO, ß-hydroxy-4-morpholine propanesulfonic acid; PIPES, 1,4-piperazine(ethanesulfonic acid); TAPS, 3-tris(hydroxymethyl)-methylamino-1-propanesulfonic acid; TES, 2-tris(hydroxymethyl)-methylamino-1-propanesulfonic acid; TRICINE, *N*-tris(hydroxymethyl)methylglycine; TRIS, tris(hydroxymethyl)-aminomethane. The pK_a values of Good's buffers are from ref *11*.

[b] pK_{a_4}, 10.6.

Appendix F

Table F-1. Useful Data for Nucleic Acids[a]

Nucleotide	Anhydrous Mol Wt[b] (free acid form)	λ_{max} (nM)	$\epsilon^{\lambda max}$ (pH 7)	$\epsilon_{260\ nm}$ (pH 7)
Adenosine triphosphate	507	259	15.4	15.0
Guanosine triphosphate	523	253	13.7	11.8
Cytidine triphosphate	483	271	9.1	7.4
Thymidine triphosphate	498	267	9.6	8.4
Uridine triphosphate	484	260	10.0	9.0
Inosine triphosphate	508	249	12.2	7.4

NOTES: Approximate mol wt of single-stranded DNA or RNA ≈ number of nucleotides × 325. Approximate mol wt of double-stranded DNA or RNA ≈ number of nucleotides × 650. A_{260}/A_{280} is 1.8 for pure DNA, and 2.0 for pure RNA. Lower values indicate contamination. One absorbance unit is 50 µg/mL for double-stranded DNA, 40 µg/mL for single-stranded DNA or RNA, 33µ g/mL for oligonucleotides with 40–100 bases, and 25 µg/mL for oligonucleotides less than 40 bases long.

[a] See also Table 3.1.

[b] To obtain the molecular weight of the diphosphates, monophospates, and unphosphorylated, subtract 80 from each missing phosphate. The spectral data given apply to the phosphorylated forms also, except in occasional cases where λ_{max} differs slightly.

APPENDIX G

Table G-1. Symbols, Molecular Weight, and pK_a Values of α-Amino Acids

Amino Acid	3-Letter Symbol	1-Letter Symbol	Mol Wt	pK_a of:		
				α-COOH	α-NH_3^+-	–R group
Alanine	Ala	A	89	2.35	9.69	—
Arginine	Arg	R	174	2.17	9.04	12.48
Asparagine	Asn	N	132	2.02	8.8	—
Aspartic acid	Asp	D	133	2.09	9.82	3.86
Cysteine	Cys	C	121	1.71	10.78	8.33
Glutamic acid	Glu	E	147	2.19	9.67	4.25
Glutamine	Gln	Q	146	2.17	9.13	—
Glycine	Gly	G	75	2.34	9.6	—
Histidine	His	H	155	1.82	9.17	6.0
Isoleucine	Ile	I	131	2.36	9.68	—
Leucine	Leu	L	131	2.36	9.60	—
Lysine	Lys	K	146	2.18	8.95	10.53
Methionine	Met	M	149	2.28	9.21	—
Phenylalanine	Phe	F	165	1.83	9.13	—
Proline	Pro	P	115	1.99	10.60	—
Serine	Ser	S	105	2.21	9.15	—
Threonine	Thr	T	119	2.63	10.43	—
Tryptophan	Trp	W	204	2.38	9.36	—
Tyrosine	Tyr	Y	181	2.20	9.11	10.07
Valine	Val	V	117	2.32	9.62	—

Appendix H

Recipes for Preparing Common Laboratory Buffers

1. Phosphate Buffer

To prepare phosphate buffer at any concentration (*X* molar), first prepare *X* molar solutions of: (1) HPO_4^{2-} using K_2HPO_4 or Na_2HPO_4, and (2) $H_2PO_4^-$ using KH_2PO_4 or NaH_2PO_4. Prepare 100 mL, 300 mL, and 1 L of the buffer, following the table below. Check the pH with a pH electrode: if a slight adjustment is needed, add concentrated acid or base dropwise.

Table H-1. Phosphate Buffer

	To Prepare					
	100 mL Buffer		300 mL Buffer		1000 mL Buffer	
pH	Add This Amount of *X* Molar HPO_4^{2-} (mL)	To this Amount of *X* Molar $H_2PO_4^-$ (mL)	Add this Amount of *X* Molar HPO_4^{2-} (mL)	To this Amount of *X* Molar $H_2PO_4^-$ (mL)	Add this Amount of *X* Molar HPO_4^{2-} (mL)	To this Amount of *X* Molar $H_2PO_4^-$ (mL)
6.3	24.0	76.0	72.0	228.0	240.0	760.0
6.5	33.4	66.6	100.2	199.8	333.9	666.1
6.7	44.3	55.7	132.8	167.2	442.7	557.3
6.8	50.0	50.0	150.0	150.0	500.0	500.0
6.9	55.7	44.3	167.2	132.8	557.3	442.7
7.0	61.3	38.7	183.9	116.1	613.1	386.9
7.2	71.5	28.5	214.6	85.4	715.3	284.7
7.4	79.9	20.1	239.8	60.2	799.2	200.8
7.5	83.4	16.6	250.1	49.9	833.7	166.3
7.8	90.9	9.1	272.7	27.3	909.1	90.9

Note: [HPO_4^{2-}] and [$H_2PO_4^-$] must be identical and equal to that of the buffer to be prepared. See problem 24d.

2. Tris-HCl Buffer

To prepare Tris-HCl buffer at any concentration (*X* molar), first prepare *X* molar solutions of: (1) Tris base (2) HCl. Prepare 100 mL, 300 mL, and 1 L of the buffer following Table H-2. Check the pH with a pH electrode. If a slight adjustment is needed, add concentrated acid or base dropwise.

Table H-2. Tris-HCl Buffer

	To Prepare					
	100 mL Buffer		**300 mL Buffer**		**1000 mL Buffer**	
pH	Add This Amount of *X* Molar Tris base (mL)	To this Amount of *X* Molar HCl (mL)	Add this Amount of *X* Molar Tris Base (mL)	To this Amount of *X* Molar HCl (mL)	Add this Amount of *X* Molar Tris base (mL)	To this Amount of *X* Molar HCl (mL)
7.4	16.6	83.4	49.9	250.1	166.3	833.7
7.5	20.1	79.9	60.2	239.8	200.8	799.2
7.6	24.0	76.0	72.1	227.9	240.3	759.7
7.7	28.5	71.5	85.4	214.6	284.7	715.3
7.8	33.4	66.6	100.2	199.8	333.9	666.1
7.9	38.7	61.3	116.1	183.9	386.9	613.1
8.0	44.3	55.7	132.8	167.2	442.7	557.3
8.1	50.0	50.0	150.0	150.0	500.0	500.0
8.2	55.7	44.3	167.2	132.8	557.3	442.7
8.3	61.3	38.7	183.9	116.1	613.1	386.9
8.4	66.6	33.4	199.8	100.2	666.1	333.9
8.5	71.5	28.5	214.6	85.4	715.3	284.7
8.6	76.0	24.0	227.9	72.1	759.7	240.3
8.8	83.4	16.6	250.1	49.9	833.7	166.3

Note: [Tris base] in solution 1, nd HCl in solution 2 must be identical and equal to that of the buffer to be prepared. See problem 24.

3. Acetate Buffer

To prepare acetate buffer at any concentration (*X* molar), first prepare *X* molar solutions of: (1) acetate using sodium or potassium acetate, and (2) acetic acid. Prepare 100 mL, 300 mL, and 1 L of the buffer following the table below. Check the pH with a pH electrode. If a slight adjustment is needed, add concentrated acid or base dropwise.

Table H-3. Acetate Buffer

pH	To Prepare					
	100 mL Buffer		300 mL Buffer		1000 mL Buffer	
	Add This Amount of *X* Molar Acetate (mL)	To this Amount of *X* Molar Acetic Acid (mL)	Add this Amount of *X* Molar Acetate (mL)	To this Amount of *X* Molar Acetic Acid (mL)	Add this Amount of *X* Molar Acetate (mL)	To this Amount of *X* Molar Acetic Acid (mL)
4.0	13.7	86.3	41.0	259.0	136.8	863.2
4.2	20.1	79.9	60.2	239.8	200.8	799.2
4.4	28.5	71.5	85.4	214.6	284.7	715.3
4.5	33.4	66.6	100.2	199.8	333.9	666.1
4.6	38.7	61.3	116.1	183.9	386.9	613.1
4.7	44.3	55.7	132.8	167.2	442.7	557.3
4.8	50.0	50.0	150.0	150.0	500.0	500.0
4.9	55.7	44.3	167.2	132.8	557.3	442.7
5.0	61.3	38.7	183.9	116.1	613.1	386.9
5.1	66.6	33.4	199.8	100.2	666.1	333.9
5.2	71.5	28.5	214.6	85.4	715.3	284.7
5.3	76.0	24.0	227.9	72.1	759.7	240.3
5.4	79.9	20.1	239.8	60.2	799.2	200.8
5.6	86.3	13.7	259.0	41.0	863.2	136.8

Note: [Acetate] in solution 1 and [acetic acid] in solution 2 must be identical and equal to that of the buffer to be prepared. See problem 24.

4. Common Laboratory Buffers

Table H-4. Common Laboratory Buffers

Buffer	Concn of 1× Solution	Recipe and Instructions
Tris-EDTA	10 mM Tris-Cl 1 mM EDTA	**1× solution:** 1.21 g Tris base, 2 mL of 0.5 M EDTA. **10× solution:** 12.11 g Tris base, 20 mL of 0.5 M EDTA.
Tris-Acetate-EDTA (pH 7.8)	40 mM Tris-Acetate 1 mM EDTA	**1× solution:** 4.84 g Tris base, 1.53 mL conc. glacial acetic acid, 2 mL of 0.5 M EDTA. **10× solution:** 48.44 g Tris base, 15.29 mL conc. glacial acetic acid, 2 mL of 0.5 M EDTA. Dissolve Tris in 800 mL deionized water. While stirring, add EDTA and acetic acid (slowly). Check the pH, and if minor adjustment is needed, use acetic acid or tris base. Qs to 1L.
Tris-Borate-EDTA (pH 8.3)	90 mM Tris-Borate 1 mM EDTA	**0.5×:** 5.45 g Tris base, 3.41 g boric acid, 2 mL of 0.5 M EDTA. **1×:** 10.9 g Tris base, 6.82 g boric acid, 2 mL of 0.5 M EDTA. **10×:** 109 g Tris base, 68.2 g boric acid, 20 mL 0.5 M EDTA. Dissolve Tris and boric acid in 800 mL water. Add EDTA. Check the pH. Adjust with boric acid or tris base if necessary. Qs to 1L.

Table H-4, continued. Common Laboratory Buffers

Buffer	Concn of 1× Solution	Concn of 1× Solution
Tris-Glycine SDS (pH 8.3)	25 mM Tris 203 mM glycine 0.1% sodium dodecyl sulfate (SDS)	**1× solution:** 3.03 g Tris base, 15.2 g glycine. **10× solution:** 30.3 g Tris base, 152.0 g glycine. Dissolve Tris and glycince in 800 mL deionized water. Check the pH, and if minor adjustment is needed, use Tris base or glycine. Add and dissolve the SDS. Qs to 1 L.
Sample Buffer for PAGE and SDS-PAGE (pH 6.8)	62.3 mM Tris base 10% glycerol, 2% SDS, 140 mM ß-mercaptoethanol, 0.01% bromphenol blue.	**Amounts given per 100 mL solution.** **2×:** 1.51 g Tris base, 20 mL glycerol, 4 g SDS, 2 mL ß-mercaptoethanol, 0.02 g bromphenol blue. **5×:** 3.78 g Tris base, 50 mL glycerol, 10 g SDS, 5 mL ß-mercaptoethanol, 0.05 g bromphenol blue. Dissolve Tris in 30 mL deionozed water, add and dissolve the glycerol. Adjust the pH to 6.8 with conc. HCl. Add and dissolve the SDS, -mercaptoethanol, bromphenol blue, and qs to 1 L . Store at 25°C.

Appendix I

Table I-1. Half-Life of Selected Isotopes[a]

Isotope	Half-Life[b]	Type of Particle Emitted	Major Decay Energies (MeV)
Carbon-14 (^{14}C)	5760 years	β^-	0.155
Hydrogen-3 (^{3}H)	12.26 years	β^-	0.018
Iodine-125 (^{125}I)	60 days	EC	
Iodine-131 (^{131}I)	8.04 days	β^- γ	0.61, 0.33 0.36, 0.64
Phosphorus-32 (^{32}P)	14.2 days	β^-	1.71
Phosphorus-33 (^{33}P)	25 days	β^-	0.25
Potassium-40 (^{40}K)	1.3 x 10^9 years	β^- γ EC	1.32 1.46 —
Rubidium-86 (^{86}Rb)	18.7 days	β^- γ	1.82 1.1
Sodium-22 (^{22}Na)	2.6 years	β^+ γ	0.55, 0.58, 1.8 0.51, 1.27
Sulfur-35 (^{35}S)	87.2 days	β^-	0.167

[a] See page 135 for radioactivity units.

[b] See Source; ref 12.

Appendix J

Table J-1. Fractions of ^{32}P and ^{131}I Remaining Over Time

Time Elapsed (Days)	Radioactivity Remaining (Fraction)	
	^{32}P	^{131}I
0.5	0.976	0.958
1.0	0.953	0.918
1.5	0.930	0.880
2.0	0.908	0.843
2.5	0.886	0.807
3.0	0.865	0.774
3.5	0.844	0.741
4.0	0.824	0.710
4.5	0.804	0.681
5.0	0.785	0.652
5.5	0.766	0.625
6.0	0.748	0.599
6.5	0.730	0.573
7.0	0.712	0.549
7.5	0.695	0.526
8.0	0.679	0.504
9.0	0.647	0.463
10.0	0.616	0.425
11.0	0.587	0.390
12.0	0.559	0.358
13.0	0.533	0.329
14.0	0.507	0.302
15.0	0.483	0.277
16.0	0.461	0.254

Table J-2. Fractions of ^{125}I, and ^{35}S, Remaining Over Time

Time Elapsed (Days)	Radioactivity Remaining (Fraction)	
	^{125}I	^{35}S
2	0.977	0.984
4	0.955	0.969
6	0.933	0.953
8	0.912	0.938
10	0.891	0.923
12	0.871	0.909
14	0.851	0.895
16	0.831	0.881
18	0.812	0.869
20	0.794	0.853
22	0.776	0.839
24	0.758	0.826
26	0.741	0.813
28	0.724	0.800
30	0.707	0.788
32	0.691	0.775
34	0.675	0.763
36	0.660	0.751
38	0.645	0.739
40	0.630	0.727
42	0.616	0.716
44	0.602	0.705
46	0.588	0.694
48	0.574	0.683
50	0.561	0.672
52	0.549	0.661
54	0.536	0.651
56	0.524	0.641
60	0.500	0.620
64	0.478	0.601
68	0.456	0.582
72	0.435	0.564
76	0.416	0.546
80	0.397	0.529
84	0.379	0.513

Appendix K

Table K-1. Atomic Weights of the Selected Elements

Element (Symbol)	Atomic Weight	Element	Atomic Weight
Aluminum (Al)	26.982	Mercury (Hg)	200.59
Antimony (Sb)	121.75	Molybdenum (Mo)	95.94
Argon (Ar)	39.95	Neon (Ne)	20.183
Arsenic (As)	74.922	Nickel (Ni)	158.71
Barium (Ba)	137.34	Niobium (Nb)	92.906
Beryllium	9.012	Nitrogen (N)	14.0067
Bismuth (Bi)	208.980	Osmium (Os)	190.2
Boron (B)	79.909	Oxygen (O)	15.999
Cadmium (Cd)	112.40	Palladium (Pd)	106.4
Calcium (Ca)	40.08	Phosphorus (P)	30.9738
Carbon (C)	12.011	Platinum (pt)	195.09
Cesium (Cs)	132.905	Potassium (K)	39.102
Chlorine (Cl)	35.453	Radon (Rn)	222
Chromium (Cr)	51.996	Rhodium (Rh)	102.905
Cobalt (Co)	58.933	Rubidium (Rb)	85.47
Copper (Cu)	63.54	Selenium (Se)	78.96
Fluorine (Fl)	18.998	Silicon (Si)	28.086
Gold (Au)	196.967	Silver (Ag)	107.78
Helium (He)	4.003	Sodium (Na)	22.9898
Hydrogen (H)	1.008	Sulfur (S)	32.064
Indium (In)	114.82	Thallium (Ti)	204.37
Iodine (I)	126.904	Tin (Sn)	118.69
Iridium (Ir)	192.2	Titanium (Ti)	47.90
Iron (Fe)	55.847	Tungsten (W)	183.85
Lanthanum (La)	138.91	Uranium (U)	238.03
Lead (Pb)	207.19	Vanadium (V)	50.942
Lithium (Li)	6.939	Xenon (Xe)	131.30
Magnesium (Mg)	34.312	Ytterbium (Yb)	173.04
Manganese (Mn)	54.938	Zinc (Zn)	65.37

Index to Practical Examples

Page numbers are enclosed in parentheses.
A problem may appear under one or more headings.

Radioactivity

General Index

S

T – Z